# Assessment of Agent Monitoring Strategies for the Blue Grass and Pueblo Chemical Agent Destruction Pilot Plants

Committee on Assessment of Agent Monitoring Strategies for the Blue Grass and Pueblo Chemical Agent Destruction Pilot Plants

Board on Army Science and Technology

Division on Engineering and Physical Sciences

# NATIONAL RESEARCH COUNCIL
*OF THE NATIONAL ACADEMIES*

THE NATIONAL ACADEMIES PRESS
Washington, D.C.
**www.nap.edu**

THE NATIONAL ACADEMIES PRESS   500 FIFTH STREET, NW   Washington, DC 20001

NOTICE: The project that is the subject of this report was approved by the Governing Board of the National Research Council, whose members are drawn from the councils of the National Academy of Sciences, the National Academy of Engineering, and the Institute of Medicine. The members of the committee responsible for the report were chosen for their special competences and with regard for appropriate balance.

This study was supported by Contract No. W911NF-11-C-0033 between the National Academy of Sciences and the U.S. Army. Any opinions, findings, conclusions, or recommendations expressed in this publication are those of the author(s) and do not necessarily reflect the views of the organizations or agencies that provided support for the project.

International Standard Book Number-13: 978-0-309-25985-9
International Standard Book Number-10: 0-309-25985-1

Limited copies of this report are available from Board on Army Science and Technology, National Research Council, 500 fifth Street, NW, Room 940, Washington, DC 20001; (202) 334-3118.

Additional copies of this report are available from the National Academies Press, 500 Fifth Street, NW, Keck 360, Washington, DC 20001; (800) 624-6242 or (202) 334-3313; http://www.nap.edu.

Copyright 2012 by the National Academy of Sciences. All rights reserved.

Printed in the United States of America

# THE NATIONAL ACADEMIES
*Advisers to the Nation on Science, Engineering, and Medicine*

The **National Academy of Sciences** is a private, nonprofit, self-perpetuating society of distinguished scholars engaged in scientific and engineering research, dedicated to the furtherance of science and technology and to their use for the general welfare. Upon the authority of the charter granted to it by the Congress in 1863, the Academy has a mandate that requires it to advise the federal government on scientific and technical matters. Dr. Ralph J. Cicerone is president of the National Academy of Sciences.

The **National Academy of Engineering** was established in 1964, under the charter of the National Academy of Sciences, as a parallel organization of outstanding engineers. It is autonomous in its administration and in the selection of its members, sharing with the National Academy of Sciences the responsibility for advising the federal government. The National Academy of Engineering also sponsors engineering programs aimed at meeting national needs, encourages education and research, and recognizes the superior achievements of engineers. Dr. Charles M. Vest is president of the National Academy of Engineering.

The **Institute of Medicine** was established in 1970 by the National Academy of Sciences to secure the services of eminent members of appropriate professions in the examination of policy matters pertaining to the health of the public. The Institute acts under the responsibility given to the National Academy of Sciences by its congressional charter to be an adviser to the federal government and, upon its own initiative, to identify issues of medical care, research, and education. Dr. Harvey V. Fineberg is president of the Institute of Medicine.

The **National Research Council** was organized by the National Academy of Sciences in 1916 to associate the broad community of science and technology with the Academy's purposes of furthering knowledge and advising the federal government. Functioning in accordance with general policies determined by the Academy, the Council has become the principal operating agency of both the National Academy of Sciences and the National Academy of Engineering in providing services to the government, the public, and the scientific and engineering communities. The Council is administered jointly by both Academies and the Institute of Medicine. Dr. Ralph J. Cicerone and Dr. Charles M. Vest are chair and vice chair, respectively, of the National Research Council

**www.national-academies.org**

## COMMITTEE ON ASSESSMENT OF AGENT MONITORING STRATEGIES FOR THE BLUE GRASS AND PUEBLO CHEMICAL AGENT DESTRUCTION PILOT PLANTS

CHARLES E. KOLB, *Chair*, Aerodyne Research, Inc., Billerica, Massachusetts
JESSE L. BEAUCHAMP (NAS), California Institute of Technology, Pasadena
ROBERT A. BEAUDET, University of Southern California, Pasadena
JOAN B. BERKOWITZ, Farkas Berkowitz and Company, Washington, D.C.
HAO CHEN, Ohio University, Athens
ADRIENNE T. COOPER, Florida Agricultural and Mechnical University, Tallahassee
FACUNDO M. FERNANDEZ, Georgia Institute of Technology, Atlanta
ROBERT D. GIBBONS (IOM), University of Chicago
JOHN A. McLEAN, Vanderbilt University, Nashville, Tennessee
MAX D. MORRIS, Iowa State University, Ames
DONALD W. MURPHY (NAE), Bell Laboratories, Lucent Technologies (retired), Davis, California
C. SHANE REESE, Brigham Young University, Mapleton, Utah
LORENZ R. RHOMBERG, Gradient, Cambridge, Massachusetts
ALBERT A. VIGGIANO, Air Force Research Laboratory, Kirtland AFB, New Mexico

**Staff**

HARRISON T. PANNELLA, Study Director
NIA D. JOHNSON, Senior Research Associate
ANN F. LARROW, Research Assistant

# BOARD ON ARMY SCIENCE AND TECHNOLOGY

ALAN H. EPSTEIN, *Chair,* Pratt & Whitney, East Hartford, Connecticut
DAVID M. MADDOX, *Vice Chair,* Independent Consultant, Arlington, Virginia
DUANE ADAMS, Independent Consultant, Carnegie Mellon University (retired), Arlington, Virginia
ILESANMI ADESIDA, University of Illinois at Urbana-Champaign
MARY E. BOYCE, Massachusetts Institute of Technology, Cambridge
EDWARD C. BRADY, Strategic Perspectives, Inc., Fort Lauderdale, Florida
W. PETER CHERRY, Independent Consultant, Ann Arbor, Michigan
EARL H. DOWELL, Duke University, Durham, North Carolina
JULIA D. ERDLEY, Pennsylvania State University, State College
LESTER A. FOSTER, Electronic Warfare Associates, Herndon, Virginia
JAMES A. FREEBERSYSER, BBN Technology, St. Louis Park, Minnesota
RONALD P. FUCHS, Independent Consultant, Seattle, Washington
W. HARVEY GRAY, Independent Consultant, Oak Ridge, Tennessee
JOHN J. HAMMOND, Lockheed Martin Corporation (retired), Fairfax, Virginia
RANDALL W. HILL, JR., University of Southern California Institute for Creative Technologies, Playa Vista
JOHN W. HUTCHINSON, Harvard University, Cambridge, Massachusetts
MARY JANE IRWIN, Pennsylvania State University, University Park
ROBIN L. KEESEE, Independent Consultant, Fairfax, Virginia
ELLIOT D. KIEFF, Channing Laboratory, Harvard University, Boston, Massachusetts
WILLIAM L. MELVIN, Georgia Tech Research Institute, Smyrna
ROBIN MURPHY, Texas A&M University, College Station
SCOTT PARAZYNSKI, University of Texas Medical Branch, Galveston
RICHARD R. PAUL, Independent Consultant, Bellevue, Washington
JEAN D. REED, Independent Consultant, Arlington, Virginia
LEON E. SALOMON, Independent Consultant, Gulfport, Florida
JONATHAN M. SMITH, University of Pennsylvania, Philadelphia
MARK J.T. SMITH, Purdue University, West Lafayette, Indiana
MICHAEL A. STROSCIO, University of Illinois, Chicago
DAVID A. TIRRELL, California Institute of Technology, Pasadena
JOSEPH YAKOVAC, President, JVM LLC, Hampton, Virginia

**Staff**

BRUCE A. BRAUN, Director
CHRIS JONES, Financial Manager
DEANNA P. SPARGER, Program Administrative Coordinator

# Preface

More than 25 years ago, in 1986, the U.S. Army began destruction of its nearly 30,000-ton legacy of stockpiled chemical agents, stored in approximately 3 million individual munitions as well as numerous bulk agent containers. The nation's chemical weapons demilitarization effort has succeeded in destroying the chemical munitions and bulk agent stored at six of the eight chemical agent depots located in the continental United States. Chemical weapons that had been deployed abroad and relocated to a storage depot on Johnston Atoll, southwest of Hawaii, have also been successfully destroyed. To date, 90 percent of the original U.S. stockpile has been safely destroyed.

Six of the eight continental chemical stockpiles, as well as the Johnson Atoll site, contained large numbers of assembled chemical weapons as well as bulk agent containers, while the other two continental sites stored only bulk agent containers. The demilitarization facilities that successfully dealt with both assembled weapons and bulk agent at five storage sites used several types of specialized furnaces to incinerate chemical agent and energetic materials and decontaminate metal munitions casings, bulk agent containers, and many agent-contaminated secondary waste streams. The two demilitarization facilities dealing only with bulk agent used chemical neutralization (aqueous-based hydrolysis) reactions to fragment and detoxify the chemical agents and a combination of decontamination solutions and steam to clean the agent containers.

Demilitarization plants for the two remaining chemical weapons depots, which contain the remaining 10 percent of the nation's chemical agent in assembled chemical projectiles and rockets, are currently under construction. These facilities are funded separately under the DOD's Assembled Chemical Weapons Assessment (ACWA) program and implemented by a dedicated U.S. Army Element. Local concerns about incineration of chemical weapons forced the Army to design these facilities without the large furnaces used at other assembled chemical weapons demilitarization plants to destroy agent and energetics and to decontaminate many secondary waste materials. The lack of high-throughput furnaces to destroy or decontaminate secondary waste materials creates a need to easily and reliably determine which waste materials are contaminated with agent and if initial decontamination efforts have succeeded. Demilitarization facility closure activities might also be expedited if tools, equipment, and building surfaces could be monitored easily and reliably for agent contamination.

While the Army has developed and successfully used methods to detect chemical agent contamination of various materials, these tend to be indirect and time consuming. Recent advances in analytical instrumentation suggest that it may be feasible to deploy robust portable instruments that can detect and characterize chemical agent contamination of a wide variety of materials in real time. Formed under the auspices of the Board on Army Science and Technology (BAST), the Committee on Assessment of Agent Monitoring Strategies for the Blue Grass and Pueblo Chemical Agent Destruction Pilot Plants (ACWA Monitoring Committee) was appointed by the National Research Council to survey the capabilities of newly available analytical instrumentation, and to assess how such capabilities might be deployed to better characterize chemical agent contamination of secondary waste materials during agent destruction operations and provide a real-time monitoring tool for contaminated equipment and construction materials during closure activities at the last two U.S. chemical weapons stockpile demilitarization facilities.

In the present report, the ACWA Monitoring Committee presents its findings and recommendations to the Program Manager for Assembled Chemical Weapons Alternatives (PMACWA), whose staff is responsible for the construction, operation, and closure of the last two U.S. chemical weapons stockpile demilitarization facilities. During its deliberations, the committee benefited from the insights and analyses of senior ACWA personnel and wishes to specifically acknowledge detailed inputs about anticipated ACWA operational procedures and requirements from C.J. Anderson and J.M. Kiley. The committee also benefited greatly from the efforts of BAST's professional staff, including the study director, Harrison T. Pannella, senior research associate Nia D. Johnson, and research assistant and logistics expert Ann F. Larrow.

Charles E. Kolb, *Chair*
Committee on Assessment of Agent
    Monitoring Strategies for the Blue Grass
    and Pueblo Chemical Agent Destruction
    Pilot Plants

# Acknowledgment of Reviewers

This report has been reviewed in draft form by individuals chosen for their diverse perspectives and technical expertise, in accordance with procedures approved by the National Research Council's (NRC's) Report Review Committee. The purpose of this independent review is to provide candid and critical comments that will assist the institution in making its published report as sound as possible and to ensure that the report meets institutional standards for objectivity, evidence, and responsiveness to the study charge. The review comments and draft manuscript remain confidential to protect the integrity of the deliberative process. We wish to thank the following individuals for their review of this report:

Charles K. Bayne, Consultant;
John I. Brauman, NAS, Stanford University;
Robert B. Cody, JEOL USA, Inc.;
R. Graham Cooks, Purdue University;
Gary S. Groenewold, Idaho National Laboratory;
M. Douglas LeVan, Vanderbilt University;
Fred W. McLafferty, NAS, Cornell University;
W. Leigh Short, Consultant (retired); and
G. Geoffrey Vining, Virginia Tech.

Although the reviewers listed above have provided many constructive comments and suggestions, they were not asked to endorse the conclusions or recommendations nor did they see the final draft of the report before its release. The review of this report was overseen by Hyla S. Napadensky, Napadensky Energetics Inc. (retired). Appointed by the NRC, she was responsible for making certain that an independent examination of this report was carried out in accordance with institutional procedures and that all review comments were carefully considered. Responsibility for the final content of this report rests entirely with the authoring committee and the institution.

# Contents

EXECUTIVE SUMMARY     1

1    INTRODUCTION     5
     Report Motivation, 5
     Committee Composition, 7
     Committee Statement of Task, 7
     Committee Activities, 8
     Report Road Map, 9

2    BGCAPP AND PCAPP DESIGNS AND RELEVANT PROCEDURES
     USED AT DESTRUCTION FACILITIES     11
     Background on Safety Procedures and Requirements Used at
         Chemical Agent Destruction Facilities, 12
     Pueblo Chemical Agent Destruction Pilot Plant, 19
     Blue Grass Chemical Agent Destruction Pilot Plant, 24
         M55 Rocket Processing, 24
         Projectile Processing, 29
     Description of the HVAC Systems Used at Both Facilities, 29

3    AGENT MONITORING PRACTICES FOR WASTE GENERATED AT
     BGCAPP AND PCAPP     35
     Waste Analysis Overview, 35
     Waste Generation and Monitoring Overview, 36
     Monitoring Based on Vapor Measurements, 43
         Air Monitoring Instrumentation and Methods, 43
         Extractive Analysis, 44
     Use of DPE Suits During Plant Operations, 45
     Changeover of Agent Disposal Campaigns at BGCAPP, 50
     Closure Operations, 51
     Activated Carbon Disposal, 56
     Scenarios Summary, 57

4     CURRENT STATUS OF SURFACE MEASUREMENT TECHNOLOGIES AND POTENTIAL ACWA SITE APPLICATIONS     59
Introduction, 59
Properties of the Target Molecules Relevant to Their Detection by Ambient Mass Spectrometry, 62
Experimental Methods for Ambient Mass Spectrometry, 68
    Direct Analysis in Real Time (DART), 70
    Desorption Electrospray Ionization (DESI), 80
    Nonproximate Analysis by Ambient Mass Spectrometry, 82
Potential Roles for Ambient Mass Spectrometry in the ACWA Program, 87
Findings and Recommendations, 90

5     STATISTICAL METHODS AND MEASUREMENT     95
Overview, 95
Review of Existing Agent Measurement Approaches, 97
Analytical Measurement Issues, 98
    Calibration Designs, 99
    CWA Data Analysis Example, 100
Compliance Monitoring, 103
Statistical Sampling Issues, 104
    Measurement Bias, Precision, and Detection Limits, 104
    Measurement Basis, 105
    Spatial Modeling, 106
    Sampling Plans for Spatial Modeling, 108
    Hot Spot Detection, 110
    Sampling Plans for Hot Spot Detection, 111

6     REPORT SUMMATION AND RECOMMENDATIONS     113
Findings and Recommendations, 114
Conclusions, 123

REFERENCES     125

APPENDIXES

A     Biographical Sketches of Committee Members     137
B     Committee Meetings     143
C     Commercial Sources of Ambient Ionization Mass Spectrometry Instrumentation     147
D     Statistical Calibration     151
E     Sampling Variability and Uncertainty Analyses     163

# Tables, Figures, and Boxes

**TABLES**

2-1  Processes and Unit Operations Being Used at PCAPP and BGCAPP, 13
2-2  Airborne Exposure Limits, Vapor Screening Levels, and Acute Exposure Guideline Levels for Chemical Agents, 14
2-3  Release Levels, Based on AEL Values, for Reuse of Items, 15
2-4  Room Contamination Requirements Using Near-Real-Time Monitoring of the Vapors in the Room, 18
2-5  Chemical Weapons Stockpile of HD- or HT-Filled Munitions at Pueblo Chemical Depot, 20
2-6  Description of the Chemical Weapons in the BGAD Stockpile, 25

3-1  Projected Amounts of Mustard-Agent-Contaminated Secondary Waste from Normal Operations at PCAPP, 38
3-2  Projected Amounts of Mustard-Agent-Contaminated Secondary Waste from Closure at PCAPP, 39
3-3  Projected Secondary Waste Streams for >1 VSL Agent-Contaminated Waste During Operations and Closure at BGCAPP, 40
3-4  Projected Secondary Waste Streams for <1 VSL Agent-Contaminated Waste During Operations and Closure at BGCAPP, 41
3-5  Estimated Agent-Contaminated Waste Stream Summary for Operations and Closure at BGCAPP, 42
3-6  Critical Measurement Performance Criteria for Possible Scenarios, 57

4-1  Physical Properties of Chemical Warfare Agents, 66
4-2  List of Acronyms (Ordered Alphabetically) and Relevant References Describing Various Ambient Surface Sampling Techniques, 69
4-3  Capabilities and Limitations of Ambient Mass Spectrometry (DART and DESI) and Existing Vapor Monitoring (DAAMS and MINICAMS) Measurement Strategies, 89
4-4  Comparative Capabilities and Limitations of DART and DESI for Characterization of Contamination by ACWA-Relevant Chemical Agents (GB, VX, HD), 90

## FIGURES

2-1   PCAPP munitions process flow chart, 21
2-2   Process and waste stream diagram for PCAPP, 22
2-3   PCAPP site layout, 23
2-4   BGCAPP munitions process flow chart, 24
2-5   Process and waste stream diagram for BGCAPP, 26
2-6   BGCAPP site layout, 27
2-7   The nine activated carbon filter units for the MDB HVAC system, 31
2-8   Vestibule on the side of an MDB HVAC unit, 31
2-9   Schematic representation of airflow through the six filter banks that make up each MDB filter unit, 32
2-10  A filter tray, 32
2-11  Airflow path through a filter tray, 33

3-1   An overview of the analysis plan for PCAPP, 46
3-2   Workers in personal protective equipment working at a chemical weapons disposal facility, 48
3-3   An example of a large item tented for monitoring at closure, 54

4-1   Schematic diagram of DART ion source, 61
4-2   Schematic diagram of DESI ion source, 61
4-3   Schematic illustrations showing the operation of several different ion sources and sampling schemes for ambient mass spectrometry, 71
4-4   Additional illustrations showing the operation of several different ion sources and sampling schemes for ambient mass spectrometry, 72
4-5   Laser-based ambient ionization techniques: (left) Laser Ablation-Electrospray Ionization (LAESI) and (right) Infrared Laser Ablation Metastable-induced Chemical Ionization (IR-LAMICI), 73
4-6   Schematic illustrations (this page and facing page) showing the operation of several different ion sources and sampling schemes for ambient mass spectrometry, 75
4-7   DART mass spectra of agent standards, 78
4-8   Structures of HD, GA, GB, and VX, with CAS designations in brackets, 79
4-9   High-resolution mass spectra obtained by DART for 800 ng VX on aluminum, concrete, and a bird feather, 79
4-10  Surfaces of steel, rubber hose, concrete, and charcoal spiked with 10 ng GB (top row) and unspiked surfaces (bottom row), 80
4-11  DESI remote sampling techniques, 84
4-12  DESI remote sampling techniques using AE and AFAI, 85
4-13  RASTIR and ND EESI schematics, 85
4-14  Multiple-sprayer and nonproximate large-area sprayer DESI setups, 86

5-1   Estimated calibration function for VX in DI (deionized) water, 101

5-2 Estimated relationship between variability and concentration for VX in DI (deionized) water, 101

5-3 Relationship between the percent relative standard deviation (%RSD) and concentration for VX in DI (deionized) water, 102

## BOXES

2-1 Definition of Generator Knowledge, 33

3-1 Scenario 3A: Improving Worker Safety During DPE Entries, 49
3-2 Scenario 3B: Enabling More Efficient DPE Entries, 50
3-3 Definition and Classification of Occluded Spaces, 52
3-4 Scenario 3C: Process Area Occluded Space Surveys and/or Absorbed Agent Surveys During Changeover or Closure Activities, 53
3-5 Scenario 3D: Complex Contaminated Demilitarization Machine Needs Decontamination at Agent Changeover or Closure Activities, 53
3-6 Scenario 3E: Concrete Waste Contamination Evaluation, 56
3-7 Scenario 3F: Spent Activated Carbon Contamination Evaluation, 56

# Acronyms and Abbreviations

| | |
|---|---|
| AAS | atomic absorption spectrometry |
| ACS | agent collection system |
| ACWA | Assembled Chemical Weapons Alternatives |
| AEGLS | acute exposure guideline levels |
| AEL | airborne exposure limit |
| AFAI | air flow assisted ionization |
| ALT | Acquisitions, Logistics & Technology |
| ANCDF | Anniston Chemical Agent Disposal Facility |
| ANR | agent neutralization reactor |
| ANS | agent neutralization system |
| APB | agent processing building |
| APCI | atmospheric pressure chemical ionization |
| | |
| BGAD | Blue Grass Army Depot |
| BGCAPP | Blue Grass Chemical Agent Destruction Pilot Plant |
| BRAC | base realignment and closure |
| | |
| CAM | cavity access machine |
| CDC | Centers for Disease Control and Prevention |
| CDPHE | Colorado Department of Public Health and the Environment |
| CLLE | continuous liquid-liquid extraction |
| CMA | Chemical Materials Agency (U.S. Army) |
| CWA | chemical warfare agent |
| CWC | Chemical Weapons Convention |
| | |
| DAAMS | depot area air monitoring system(s) |
| DART | direct analysis in real time |
| DESI | desorption electrospray ionization |
| DL | detection limit |
| DMMP | dimethyl methylphosphonate |
| DPE | demilitarization protective ensemble |
| | |
| EBH | energetics batch hydrolyzer |
| ECR | explosive containment room |
| ECV | explosion containment vestibule |

| | |
|---|---|
| EDT | explosive destruction technology |
| ENR | energetics neutralization reactor |
| EPA | Environmental Protection Agency |
| EQL | expected quantitation limit |
| ERB | enhanced reconfiguration building |
| ESI | electrospray ionization |
| ESSI | electrosonic spray ionization |
| | |
| FOAK | first-of-a-kind [equipment] |
| | |
| GB | a nerve agent (sarin) |
| GC-MS | gas chromatography-mass spectrometry |
| GPL | general population limit |
| | |
| H | mustard agent |
| HD | distilled mustard agent |
| HT | distilled mustard mixed with bis(2-chloroethylthioethyl) ether |
| HVAC | heating, ventilation, and air conditioning |
| | |
| ICB | immobilized cell bioreactor |
| ICP-MS | inductively coupled plasma mass spectrometry |
| IDLH | immediately dangerous to life or health |
| IP | ionization potential |
| | |
| JACADS | Johnston Atoll Chemical Agent Disposal System |
| | |
| LADESI | laser ablation/desorption electrospray ionization |
| LCL | lower confidence limit |
| LMQAP | laboratory monitoring quality assurance plan |
| LPMD | linear projectile/mortar disassembly |
| | |
| MCP | monitoring concept plan |
| MDB | munitions demilitarization building |
| MDL | method detection limit |
| MINICAMS | miniature continuous air monitoring system(s) |
| MPT | metal parts treater |
| MS | mass spectrometry; mass spectrometer |
| MSM | munitions storage magazine |
| MS/MS | tandem mass spectrometry |
| MTU | munitions treatment unit |
| MVUE | minimum variance unbiased estimate |
| MWS | munitions washout system |
| | |
| ND-EESI | neutral desorption-extractive electrospray ionization |
| NECDF | Newport Chemical Agent Disposal Facility |

| | |
|---|---|
| NRC | National Research Council |
| OLS | ordinary least squares |
| OST | occluded space team |
| PCAPP | Pueblo Chemical Agent Destruction Pilot Plant |
| PCD | Pueblo Chemical Depot |
| PCE | protective clothing and equipment |
| PMACWA | Program Manager for Assembled Chemical Weapons Alternatives |
| PPE | personal protective equipment |
| PQL | practical quantitation limit |
| PVC | polyvinyl chloride |
| RASTIR | remote analyte sampling, transport, and ionization relay |
| RCM | rocket cutting machine |
| RCRA | Resource Conservation and Recovery Act |
| RD&D | research development and demonstration |
| SCWO | supercritical water oxidation |
| SFT | shipping and firing tube |
| SPME | solid phase microextraction |
| STEL | short term exposure limit |
| T | chemical compound $((ClCH_2CH_2)_2SCH_2CH_2)_2O$ |
| TAP | toxological agent protective |
| TCLP | toxic characteristic leaching procedure |
| TOC | total organic carbon |
| TOF | time-of-flight |
| TSDF | treatment, storage and disposal facility |
| UCL | upper confidence limit |
| UPL | upper prediction limit |
| VOC | volatile organic compound |
| VSL | vapor screening level |
| VX | a nerve agent |
| WAP | waste analysis plan |
| WCL | waste control limit |
| WLS | weighted least squares |

# Executive Summary

## MOTIVATION

January 2012 saw the completion of the U.S. Army's Chemical Materials Agency's (CMA's) task to destroy 90 percent of the nation's stockpile of chemical weapons. CMA completed destruction of the chemical agents and associated munitions stored at six of eight continental U.S. storage facilities as well as chemical weapons deployed overseas, which were transported to Johnston Atoll, southwest of Hawaii, and demilitarized there. The remaining 10 percent of the nation's chemical weapons stockpile is stored at two remaining continental U.S. depots, in Lexington, Kentucky, and Pueblo, Colorado. Their destruction has been assigned to a separate U.S. Army organization, the Assembled Chemical Weapons Alternatives (ACWA) Element.

ACWA is currently constructing the last two chemical weapons disposal facilities, the Pueblo and Blue Grass Chemical Agent Destruction Pilot Plants (denoted PCAPP and BGCAPP), with weapons destruction activities scheduled to start in 2015 and 2020, respectively. ACWA is charged with destroying the mustard agent stockpile at Pueblo and the nerve and mustard agent stockpile at Blue Grass without using the multiple incinerators and furnaces used at the five CMA demilitarization plants that dealt with assembled chemical weapons—munitions containing both chemical agents and explosive/propulsive components. The two ACWA demilitarization facilities are congressionally mandated to employ noncombustion-based chemical neutralization processes to destroy chemical agents. Chapter 2 of this report reviews the disposal technologies designed to demilitarize chemical agents and other weapons components planned for use at the two ACWA facilities. ACWA will not have large furnaces to decontaminate or destroy munitions components and to process related secondary waste streams (which are cataloged in Chapter 3). This constraint has motivated an interest in analytical methods that can quickly and reliably identify and characterize agent-contaminated materials. Such methods could be useful in characterizing secondary waste materials during disposal operations at both sites, as well as during agent changeover operations (BGCAPP only) and especially during facility closure activities, when agent disposal facilities must be decontaminated before demolition (both BGCAPP and PCAPP).

In order to safely operate its disposal plants, CMA developed methods and procedures to monitor chemical agent contamination of both secondary waste materials and plant structural components. ACWA currently plans to adopt these methods and procedures (described briefly in Chapter 2 and more thoroughly in Chapter 3) for use at these facilities. While these methods have allowed safe waste processing and closure activities, they are tedious and indirect, generally relying on vapor-phase agent measurements over confined surfaces rather than direct detection of surface contamination. Chapter 3 also develops and describes a half-dozen scenarios involving prospective ACWA secondary waste characterization, process equipment maintenance and changeover activities, and closure agent decontamination challenges, where direct, real-time agent contamination measurements on surfaces or in porous bulk materials might allow more efficient and possibly safer operations if suitable analytical technology is available and affordable.

## TECHNOLOGY OPPORTUNITY

The last 5 years have produced very rapid development of ambient ionization mass spectrometric techniques capable of real-time surface and bulk material chemical analyses with little or no sample preparation. Two of these technologies, desorption electrospray ionization (DESI), first introduced in 2004, and direct analysis in real time (DART), introduced in 2005, are now commercially available and have widespread applications. Both DESI and DART, as well as a range of variations on each, are reviewed in Chapter 4, which also details their application to the detection and quantification of chemical agent and agent simulant compounds. The utility of various DESI and DART implementations to meet the chemical agent contamination characterization challenges identified in several of the ACWA operational and closure activity scenarios developed in Chapter 3 is addressed in Chapter 4. Chapter 4 also discusses the potential utility of real-time agent vapor concentration gradient measurement methods to locate specific contaminated surfaces. It concludes with a comparison of the strengths and weaknesses in (1) the current Army indirect methods to characterize chemical agent contamination adsorbed on solid surfaces and absorbed by porous materials and (2) direct measurements using ambient ionization techniques. Based on this comparison, the chapter presents findings and recommendations involving specific ambient ionization technology configurations that could guide specification, acquisition, and integration of this technology, if ACWA management determines that the ability to directly characterize the distribution of chemical agent concentrations on and in solid materials and the concentration gradients of airborne chemical agents is sufficiently useful to justify the effort and cost required to implement this type of technology.

Efficient and effective use of any analytical technology must employ statistically robust measurement strategies. This is particularly true when dealing with the characterization of contamination by deadly substances like chemical agents. Chapter 5 briefly examines the published statistical basis of current Army chemical agent contamination characterization methods, presents a detailed analysis of recently published DART measurements of chemical agent concentrations in liquid solutions, and then examines the statistical basis of direct surface agent contamination characterization using various implementations of ambient ionization mass spectrometry techniques.

## CONCLUSIONS

A compilation of all of the findings and recommendations developed in Chapters 2-5 of this report is presented in Chapter 6. As reflected in those findings and recommendations, the committee concluded that ambient ionization mass spectrometry is a rapidly maturing and highly useful technology with specific available implementations capable of highly sensitive, real-time measurements of relative concentrations of chemical agents adsorbed on a variety of relevant surfaces and in some porous materials. Further, with suitable reference standards, absolute measurements of agent concentrations in ambient air and liquid solutions are feasible. If adopted, these capabilities might be very useful in supplementing the Army's traditional air and vapor headspace agent contamination measurements using current near-real-time agent monitors. A range of scenarios occurring during agent disposal operations and facility closure activities have been defined and developed by the committee to illustrate the potential utility of real-time ambient ionization mass spectrometric detection of chemical agent contamination.

Although commercially available ambient ionization mass spectrometry instrumentation in the specific configurations recommended by the committee may not currently be available off the shelf, the major components have been commercialized, and a number of analytical instrument vendors are capable of designing, assembling, and demonstrating instruments meeting potential ACWA specifications. Given the current schedules for anticipated PCAPP and BGCAPP weapons disposal (beginning in 2015 and 2020, respectively) and facility closure activities, it is very likely that these instruments could be specified, tested, and deployed quickly enough to be used at PCAPP and BGCAPP, as suggested in this report.

In addition, as demonstrated by their work as reviewed in Chapters 4 and 5, Army scientists at the Edgewood Chemical and Biological Center, sited near ACWA headquarters, have significant experience in the application of ambient ionization mass spectrometric measurements of chemical agent concentrations and distributions and could be a resource for developing and testing specific ambient ionization technology implementations for ACWA.

Based on these considerations the committee's overarching finding and recommendation are as follows:

**Finding 6-1.** Suitably specified ambient ionization mass spectrometry instrumentation could be utilized in a range of challenging activities at ACWA chemical weapons disposal facilities where real-time chemical agent contamination measurements may reduce the time and effort required to characterize the chemical agent contamination of waste materials, process equipment, and work areas.

**Recommendation 6-1.** ACWA should carefully evaluate the capabilities of portable ambient ionization mass spectrometry and its potential to provide faster and more accurate characterization of chemical agent contamination, as detailed in this report, and determine if these likely benefits justify the effort and investment required to specify, acquire, and deploy suitable implementations of this technology.

# 1
# Introduction

**REPORT MOTIVATION**

During the middle third of the twentieth century, the United States developed and produced a wide range of weapons designed to disperse both nerve (GB and VX) and blister (sulfur mustard: H, HD, and HT) chemical agents at lethal concentrations. Although never used, 31,496 tons of chemical agents were produced after World War II, a large fraction of which was loaded into millions of individual munitions, with the rest stored in bulk containers. The agent-filled munitions and bulk containers that remained in the United States were stored in eight continental weapons depots, while chemical munitions that had been sent overseas were consolidated and stored on Johnson Atoll, southwest of the main Hawaiian islands.

In 1985, Congress directed the U.S. Army to start destruction of some elements in the chemical weapons stockpile (Public Law 99-145), and in 1991 Congress directed that all chemical weapons be destroyed (Public Law 102-484). In 1997 Congress ratified the Chemical Weapons Convention (CWC), an international treaty that specified all chemical weapons would be destroyed by April 29, 2012. What is now the Army's Chemical Materials Agency (CMA) has built and operated chemical agent disposal facilities at Johnson Atoll and six of the eight continental U.S. storage depots, successfully destroying the agent-filled munitions and/or agent in bulk containers at all seven CMA sites. These activities have successfully destroyed 90 percent of the nation's stockpiled chemical agents.

Congress assigned the job of destroying the stockpiled chemical weapons at the two remaining continental U.S. depots (in Lexington, Kentucky, and Pueblo, Colorado) to a separate U.S. Army Element, the Assembled Chemical Weapons Alternatives (ACWA) organization. This occurred after residents near these two facilities convinced their congressional representatives that they were seriously concerned about the safety and effectiveness of CMA's technology selection for assembled weapons, based on robotic disassembly and separate incineration of the chemical agent and energetic materials in agent-filled munitions. The time required to define, assess, and develop alternative assembled weapons demilitarization technologies, coupled with serious budget constraints, has delayed chemical weapons destruction at Lexington and Pueblo; their demilitarization plants are still under construction.

All chemical weapon demilitarization technologies generate large amounts of potentially agent-contaminated secondary wastes as well as contaminated equipment, machinery, and plant structural elements. Chemical agent contamination levels for secondary waste, process machinery, and demilitarization equipment (tools, respirators, instruments, etc.) resulting from agent and energetics destruction processes have to be determined in order to develop safe decontamination strategies and/or disposal options. Furthermore, demilitarization plant structural elements may need to be decontaminated for maintenance activities required during disposal operations and agent changeover breaks in plant operations and during plant closure.

CMA has developed methods for determining whether waste materials, equipment, machinery, and even structural elements are contaminated by chemical agents. These methods generally involve (1) isolating the objects of interest in some sort of enclosure and (2) after a specified equilibration time, measuring vaporized agent concentrations in the enclosure's headspace. In other cases, wipe samples of surfaces can be obtained and solvents may be used to extract chemical agent contaminants from wipe samples or waste stream materials and then analyzed with chromatographic techniques. These proven techniques are available and are planned for use in the two new ACWA demilitarization plants now being built.

While effective, these traditional methods can be time consuming and have to be repeated if initial decontamination efforts are not successful. A recent review, focused on technologies for quantifying agent vapor concentrations in CMA demilitarization facilities (NRC, 2005a), noted the potential of then newly developed ambient surface ionization mass spectrometry techniques to provide real-time measurements of chemical agent contamination on surfaces. Ambient ionization mass spectrometry involves sampling and ionization of chemical species in their native environment without, or with minimal, sample preparation. Objects sampled are usually solids, and the experiment is typically done at atmospheric pressure, with total analysis times on the order of a few seconds. Subsequent development of a wide range of ambient surface ionization techniques with mass spectrometric detection has been rapid and impressive. Significant improvements in sensitivity, response time, portability, and reliability have been demonstrated, and an increasing number of systems are or will soon be commercially available.

This study is focused on whether ambient surface ionization analytical techniques are sufficiently sensitive, specific, rapid, robust, and available to supplement current Army methods for screening materials, equipment, and structural elements for agent contamination. Considerable time and effort are spent characterizing and decontaminating secondary waste, process machinery, and equipment during both disposal operations and plant closure; this significantly extends the time needed for safe and effective chemical weapons destruction and prolongs plant closure. If robust, portable, real-time surface agent contamination analytical instrumentation is available and can shorten the time and effort required for tasks not directly contributing to chemical weapons destruction, they may be well worth deploying at the two new ACWA demilitarization plants.

## COMMITTEE COMPOSITION

The Committee on Assessment of Agent Monitoring Strategies for the Blue Grass and Pueblo Chemical Agent Destruction Pilot Plants (or ACWA Monitoring Committee) is made up of experts on analytical chemistry (including mass spectrometry and ion mobility measurements), plasma chemistry, environmental chemistry, environmental engineering, statistical sampling and experimental design, process chemical engineering, materials science, quantitative risk assessment, industrial engineering, and environmental regulations. Short biographies of the committee members are presented in Appendix A.

## COMMITTEE STATEMENT OF TASK

The committee's statement of task is as follows:

The Program Manager for Assembled Chemical Weapons Alternatives (PMACWA) is currently in the process of constructing two chemical agent destruction facilities, the Blue Grass Chemical Agent Destruction Pilot Plant (BGCAPP) in Richmond, Kentucky and the Pueblo Chemical Agent Destruction Pilot Plant (PCAPP) in Pueblo, Colorado. According to the current plant design, BGCAPP will dispose of mustard agent H, and nerve agents GB (sarin) and VX using chemical neutralization followed by supercritical water oxidation and PCAPP will dispose of mustard agent HD and HT, using chemical neutralization followed by biotreatment. In addition, the selection of an auxiliary explosive destruction technology (EDT) to handle leakers and reject munitions at PCAPP, and exploration of the use of this type of technology to possibly process M55 rocket motors and mustard agent munitions at BGCAPP is currently being investigated by ACWA.

Construction of BGCAPP and PCAPP is in the early stages and chemical agent monitoring strategies have not yet been finalized for these facilities. PMACWA will adopt monitoring methods for airborne agent at BGCAPP and PCAPP similar to those used by the U.S. Army Chemical Materials Agency (CMA) at its facilities. The current U.S. Army monitoring procedures have been considered to be sound and proven. However, PMACWA could benefit from a National Research Council (NRC) assessment of any technology advancements to monitor for agent contamination of solid waste materials as well as facility equipment and surfaces at BGCAPP and PCAPP using emerging technologies. Advancements in real-time and near real-time monitoring technology for chemical agents, agent simulants and similar semi-volatile chemicals are being regularly documented in the scientific literature and may present opportunities to build additional efficiencies into waste disposal and closure activities at BGCAPP and PCAPP.

The National Research Council will establish an ad hoc committee to:

- Review the process designs for both BGCAPP and PCAPP to evaluate the expected degree of contamination for facility solid wastes, equipment, and surfaces anticipated from agent destruction processes.

- Evaluate novel candidate technologies capable of real-time and near real-time quantification of chemical agents adsorbed onto or absorbed into materials relevant to chemical demilitarization operational and closure processes and identify specifications required for a new monitoring technology to improve BGCAPP and PCAPP waste disposal and plant closure activities.
- Using the specifications identified in bullet two, review and assess new and emerging technologies that enable rapid measurement of the degree of contamination of process equipment and waste, or if the degree of decontamination achieved is sufficient to meet established regulatory requirements, including requirements for off-site shipment of wastes. Specifically, promising novel technologies will be evaluated for detection and quantification of chemical agents adsorbed onto DPE suits, agents absorbed into activated charcoal, and agents adsorbed onto and/or absorbed into concrete surfaces.

## COMMITTEE ACTIVITIES

The committee met in Aberdeen, Maryland (February 2011), Pueblo, Colorado (June 2011), Washington, D.C. (August 2011), and Irvine, California (October, 2011). Briefings on chemical demilitarization activities and technologies and ambient ionization analytical techniques were presented by the following organizations:

- Army Element Assembled Chemical Weapons Alternatives
- Army Chemical Materials Agency
- Army Edgewood Chemical Biological Center
- Battelle Memorial Institute
- Idaho National Laboratory
- Purdue University, Department of Chemistry

At these meetings, committee members also went over technical details, discussed, outlined, and wrote draft report sections, and reviewed report drafts. Additional information on the committee's meetings is given in Appendix B.

Based on their respective expertise and experience, committee members were assigned to one or more of four working groups. These groups performed the bulk of the technical analyses and draft writing assignments. The four working groups were these:

- Plant Processes
- Current Waste and Structural Contamination Evaluation Practices
- Analytical Technologies
- Measurement Methodologies

## REPORT ROAD MAP

Chapter 2 of this report describes the planned chemical weapons destruction processes to be employed at the two ACWA demilitarization facilities and analyzes the probable secondary waste streams and planned waste treatment and disposal activities. Current Army methods for monitoring chemical agent contamination of secondary waste materials as well as equipment, machinery, and structural components during agent changeout and plant closure activities are first presented in Chapter 2 and then elaborated on in Chapter 3. Chapter 3 also presents potential scenarios in which real-time surface or condensed phase agent measurements might be useful. Chapter 4 reviews the current state of the art of analytical ambient ionization mass spectrometry and discusses potential analysis tasks and technology implementations. An analysis of statistical sampling design methods and information on their application to chemical agent surface contamination measurements in relevant chemical demilitarization plant scenarios are presented in Chapter 5. The committee's findings and recommendations are compiled in Chapter 6. As previously indicated, Appendix A contains short biographical sketches of the committee members and Appendix B provides details of committee meeting activities. Appendix C lists some known current sources of commercially available ambient ionization mass spectrometry equipment in support of the technical presentation in Chapter 4. Appendix D on statistical calibration and Appendix E on sampling variability and uncertainty analyses supplement the statistical presentation in Chapter 5.

# 2
# BGCAPP and PCAPP Designs and Relevant Procedures Used at Destruction Facilities

As described in Chapter 1, during the 2002-2003 time frame, the U.S. Department of Defense Assistant Secretary of the Army for Acquisition, Logistics, & Technology (AL&T) issued a number of directives adopting the use of neutralization (chemical hydrolysis) as the primary means for destroying the chemical agent in assembled chemical munitions stored at both the Pueblo Chemical Depot (PCD) in Colorado and the Blue Grass Army Depot (BGAD) in Kentucky.

In brief, the neutralization process entails mixing the chemical agent mustard with hot water or the nerve agents GB and VX with hot caustic solution.[1] Under these conditions, the chemical agents react with water, producing hydrolysate products that are less toxic but still require further treatment. At two former sites that only had bulk agent stored in 1-ton containers, neutralization had been used successfully and the Army was able to send the hydrolysate off-site for final treatment. In contrast, at BGCAPP and PCAPP, current plans call for on-site secondary treatment of hydrolysate before the final products can be sent off-site for disposal.

The processes selected by the Army for secondary treatment of hydrolysate are biotreatment at Pueblo and supercritical water oxidation at Blue Grass. These processes are adapted and implemented by the site system contractors and are subject to validation of their effectiveness, budgetary constraints and state regulations.

The disposal processes planned for use at PCAPP and BGCAPP and the waste streams they will produce are described below. More detailed descriptions of the unit operations can be found in prior NRC reports (NRC 2005b, 2005c) and on the ACWA Web site.[2]

At the time this report was prepared, PCAPP had been issued a Resource Conservation and Recovery Act (RCRA) Research Development and Demonstration (RD&D) permit from the Colorado Department of Public Health and the Environment (CDPHE). BGCAPP has also applied for a RCRA RD&D permit from the Kentucky Department of Environmental Protection.[3] In recognition of the significant degree to which the alternative technology designs for the two sites include first-of-a-kind (FOAK)

---

[1]Detailed information on the composition and the chemical and physical properties of the mustard agents (HD, HT) and nerve agents (GB and VX) relevant to ACWA agent destruction activities is presented in Chapter 4 (see Table 4-1 and Figure 4-8).

[2] The ACWA Web site is at http://www.pmacwa.army.mil/.

[3]Additional information concerning the PCAPP and BGCAPP permits can be found through the site links at http://www.pmacwa.army.mil/.

equipment and process technologies that have not been previously implemented at full scale, these facilities are designated as pilot plants, i.e., the Pueblo Chemical Agent Destruction Pilot Plant (PCAPP) and the Blue Grass Chemical Agent Destruction Pilot Plant (BGCAPP). However, each facility will have the capacity to destroy the entire chemical weapons stockpile stored at its location. The designs for these facilities have undergone a number of revisions from their original design proposals in 2004, and some downsizing for both technical and budgetary reasons. However, the designs have now been fixed and construction of the two plants is well under way. The RD&D permitting process recognizes the need for flexibility, and negotiations with regulators are ongoing at both sites. Table 2-1 lists the process equipment and machinery at both sites and their potential for being contaminated by agent.

In this chapter, the committee provides background on the safety regulations, procedures, and terminology that are used at chemical destruction facilities. The agent process designs at PCAPP and BGCAPP are also briefly described, followed by a description of the heating, ventilation, and air conditioning (HVAC) systems (the largest source of potentially contaminated activated carbon) common to both plants.

## BACKGROUND ON SAFETY PROCEDURES AND REQUIREMENTS USED AT CHEMICAL AGENT DESTRUCTION FACILITIES

The safety procedures and activities related to agent contamination at chemical demilitarization facilities are based on the airborne exposure limits (AELs) that were set by the Centers for Disease Control and Prevention (CDC) in 2003 and 2004 and adopted by the Army, as described in pamphlet DA PAM 385-61, *Toxic Chemical Agent Safety Standards* (U.S. Army, 2008). The draft document entitled *Assembled Chemical Weapons Alternatives (USAE ACWA) Chemical Agent Monitoring Concept Plan*, which describes in detail the standards, processes, and procedures for protecting personnel and the public at the two ACWA sites, is the primary reference for the discussion that follows (U.S. Army, 2011a). Types of AELs based on vapor concentrations and duration of inhalation exposure doses are defined and presented for the three relevant chemical agents in Table 2-2. Guidelines from Volume 3 of *Acute Exposure Guideline Levels for Selected Airborne Chemicals* that were developed as acute exposure guideline levels (AEGLs) for various chemical agents, specifically GB, VX, and mustard agent in this case, are also included in Table 2-2 (NRC, 2003). Both sets of exposure levels are used by the Army.

As briefly mentioned in Chapter 1, the contamination level of waste is often determined by sealing it in a bag or, for large equipment, a plastic "tent" enclosure at 70°F or warmer, for a time sufficient for agent vapor to equilibrate with the waste in the ambient air space. Headspace agent vapor concentrations are then determined by a near-real-time (NRT) agent monitor, such as a miniature continuous air monitoring system (MINICAMS). The Army defines NRT as a measurement cycle of between 5 and 15 minutes. This is a system that provides monitoring for airborne chemical warfare agent using an automated gas chromatograph. The vapor screening levels (VSLs) defined and presented in Table 2-2 typically determine whether the waste can be shipped off-site without further on-site treatment in accordance with site-specific RCRA permit requirements. To be reutilized, an item must meet release levels given in Table 2-3.

TABLE 2-1 Processes and Unit Operations Being Used at PCAPP and BGCAPP

| FOAK Equipment | Site(s) | Function | Note |
|---|---|---|---|
| Rocket shear machine (RSM) | BGCAPP | To separate rocket motors from the warhead, drain agent from the warhead, and shear the warhead into small pieces to be sent to the EBHs | This unit will not be contaminated unless a leaking munition contaminates it. |
| Linear projectile mortar disassembly machine (LMPD) | BGCAPP PCAPP | To disassemble projectiles and mortars and remove their bursters | This unit will not be contaminated unless a leaking munition contaminates it. |
| Munitions washout station (MWS) | BGCAPP PCAPP | To remove the burster well from projectiles, drain the chemical agent, and wash out any agent residues | This unit and the room will be contaminated. |
| Energetic bulk hydrolyser (EBH) | BGCAPP | To neutralize energetics and any chemical agent in the metal parts of the rockets and fuzes from projectiles | This unit and the explosive containment room will be contaminated. |
| Metal parts treater (MPT) | BGCAPP | To decontaminate projectile bodies and secondary waste by heating to over 1000°F for more than 15 min | The front end of this unit will be in a contaminated atmosphere, but the back end, where the metal parts are removed, will not be in a contaminated area. |
| Munitions treatment unit (MTU) | PCAPP | To decontaminate projectile bodies and secondary waste by heating to over 1000°F for more than 15 min | The front end of this unit will be in a contaminated atmosphere, but the back end, where the metal parts are removed, will not be in a contaminated area. |
| Supercritical water oxidation (SCWO) | BGCAPP | To treat agent energetics hydrolysates before releasing them for final disposal | This unit is outside the agent area fence and will be in a non-contaminated area. |
| Immobilized cell bioreactors (ICBs) | PCAPP | To treat mustard hydrolysate before releasing it for final disposal | This unit is outside the agent area fence and will be in a non-contaminated area. |

SOURCE: NRC, 2011.

Vapor screening methods are also used to determine if potentially contaminated workspaces require decontamination during agent changeover or closure activities. In these cases, NRT agent vapor monitors are deployed in unventilated areas to determine if agent vapor levels rise above 1 VSL. While these vapor screening procedures have been used to safely characterize agent contamination of both solid waste materials and structural components at CMA chemical demilitarization sites for the past two decades, they are time consuming and do not directly identify the specific contaminated surface areas that may require further decontamination.

TABLE 2-2 Airborne Exposure Limits, Vapor Screening Levels, and Acute Exposure Guideline Levels for Chemical Agents

| Exposure Limit Type | Definition | Agent-Specific Quantities (mg/m$^3$) | | |
|---|---|---|---|---|
| | | GB | VX | HD |
| **Airborne Exposure Limit** | | | | |
| General population limit (GPL) | The concentration limit in which an unprotected general population can be exposed 24 hr/day forever without any adverse effects. Time-weighted average: HD, 12 hr; VX and GB, 24 hr. | $1 \times 10^{-6}$ | $6 \times 10^{-7}$ | $2 \times 10^{-5}$ |
| Worker population limit (WPL) (8 hours) | The concentration at which an unprotected worker can operate safely 8 hr a day for 5 days per week for a working lifetime without adverse effects. Time-weighted average all agents: 8 hours/workday and 40 hour/week for 30 years. | $3 \times 10^{-5}$ | $1 \times 10^{-6}$ | $4 \times 10^{-4}$ |
| Short term exposure limit (STEL) | The level at which an unprotected worker can operate for a 15 min period. The frequency is once per day for HD and VX and four times per day for GB during an 8 hr workday. Time-weighted average: 15 min. | $1 \times 10^{-4}$ | $1 \times 10^{-5}$ | $3 \times 10^{-3}$ |
| Immediately dangerous to life or health limit (IDHL) | An atmosphere that poses an immediate threat to life, would cause irreversible adverse health effects, or would impair the ability to escape from the atmosphere. Also, the maximum level to which an unprotected worker can be exposed for 30 min without experiencing escape-impairing or irreversible health effects. | $1 \times 10^{-1}$ | $3 \times 10^{-3}$ | $7 \times 10^{-1}$ |
| **Vapor Screening Level** | | | | |
| 1 VSL | The concentration of a chemical agent in a headspace below which the materiel can be treated as uncontaminated and workers can work with only a slung protective mask. | $1 \times 10^{-4}$ | $1 \times 10^{-5}$ | $3 \times 10^{-3}$ |
| **Acute Exposure Guideline Levels**[a] | | | | |
| 1-hr AEGL-1 | | $2.8 \times 10^{-3}$ | $1.7 \times 10^{-4}$ | $6.7 \times 10^{-2}$ |
| 8-hr AEGL-1 | The airborne concentration of a substance above which it is predicted that the general population, including susceptible individuals, could experience discomfort, irritation, or certain asymptomatic nonsensory effects. However, the effects are not disabling and are transient and reversible upon cessation of exposure. | $1.0 \times 10^{-3}$ | $7.1 \times 10^{-5}$ | $8.0 \times 10^{-3}$ |
| 1-hr AEGL-2 | | $3.5 \times 10^{-2}$ | $2.9 \times 10^{-3}$ | $1.0 \times 10^{-1}$ |
| 8-hr AEGL-2 | The airborne concentration of a substance above which it is predicted that the general population, including susceptible individuals, could experience irreversible or other serious, long-lasting adverse effects or an impaired ability to escape. | $1.3 \times 10^{-2}$ | $1.0 \times 10^{-3}$ | $1.3 \times 10^{-2}$ |

[a] The acute exposure guidelines (AEGLs) are a hazard communication measure developed by the National Advisory Committee to establish acute exposure guideline levels for hazardous substances. The committee developed detailed guidelines for devising uniform, meaningful emergency response standards for the general public.

SOURCE: Adapted from U.S. Army, 2008; U.S. Army, 2011a; NRC, 2005d.

TABLE 2-3 Release Levels, Based on AEL Values, for Reuse of Items

| Classification Level[a] | Vapor Screening Level | Health-Based Risk Analysis Required[b] |
|---|---|---|
| Contaminated—Do Not Release; specific safeguards required | ≥1 VSL | No |
| Release to Agent Workers Clean—Restricted | <1 VSL | Yes |
| Release to Nonagent Workers Clean—Restricted | <1 WPL[c,d] | Yes |
| Unrestricted Release to Public (Clean—Unrestricted) | <1 GPL[d] | Yes |
| Never Contaminated—Clean | N/A | Yes |

[a]Restrictions may preclude disassembly or applying heat or friction (such as grinding) without special controls.
[b]Health-based criteria/risk analyses allow for other methods to be used or developed to determine which classification level applies.
[c]Restricted—Maintenance or disassembly of items will only be done by personnel knowledgeable in agent symptoms and characteristics, and in facilities equipped with appropriate safeguards to control potential hazards.
[d]Unrestricted—Items have been previously disassembled and are clean so that they can be released to the worker population without risk of agent release. Release must be in accordance with an approved decontamination plan.

SOURCE: Jeffrey Kiley, CMA, "Monitoring Concepts for Site Closure," presentation to the 11th International Chemical Weapons Demilitarisation Conference and Exhibition, May 2009.

It is also worth noting that the vapor screening methods described above do not directly measure the amount of chemical agent contaminating solid waste or structural surfaces. Vapor screening detects the gas phase concentrations of the relevant chemical agents, which for significant amounts of neat liquid or solid agent are controlled by the agent's vapor pressure. Vapor pressures are a function of sample temperature; values at 25°C for the chemical agents relevant to the two ACWA site stockpiles are presented in Chapter 4 (see Table 4-1). Bizzigotti et al. (2009) present temperature-dependent vapor pressure data for these and other common chemical agents. Detecting a gas-phase agent concentration equivalent to its vapor pressure in the headspace of an enclosed space indicates the presence of a reservoir of condensed-phase agent but does not determine how large that reservoir may be. Operationally, that is not an issue since the VSLs are many orders of magnitude lower than agent equilibrium vapor pressures, and so obtaining readings of sub-VSL headspace concentrations is expected to prevent release of any highly contaminated materials containing significant amounts of liquid or solid agent.

However, the measurement of agent vapor levels well below their equilibrium vapor pressures may not always indicate very low material contamination levels, because the effective vapor pressure of any condensed-phase material can be dramatically reduced if the material is dissolved in bulk liquid or surface liquid layers, if it is physically or chemically adsorbed onto solid materials surfaces, and/or if it diffuses into and binds to porous materials (such as activated carbon or concrete). Since the effective vapor pressure of a specific agent at a given temperature depends strongly on the composition, morphology, and quantity of the solid and/or liquid substrates it encounters, a given equilibrated vapor concentration measurement does not necessarily constitute even a relative measurement of the amount of agent contaminating the condensed-phase materials within the enclosed space. Adamson and Gast (1997) discuss quantitative treatments of vapor adsorption and chemisorption by well-characterized condensed phase surfaces. However, quantitative data for the interactions of chemical agent vapors with real-world, mixed material surfaces are not available.

**Finding 2-1.** The prevalent Army demilitarization activity methods of detecting materials' surface contamination involve enclosing materials and monitoring headspace agent concentrations. These are indirect methods that can determine if significant levels of agent are present in the enclosed volume; surfaces are not directly monitored. However, vapor detection does not identify the location or quantify the level of contamination on surfaces within the test volume.

The amount and spatial distributions of chemical agent contaminating secondary waste or structural materials could be determined by direct measurement of adsorbed or chemisorbed agent on their surfaces. Recent advances in analytical technology have greatly simplified the rapid and sensitive measurement of the chemical composition of solid surfaces and matrices. These advances are reviewed and discussed in depth in Chapter 4. Neither CMA nor ACWA has established a health hazard standard for surface adsorbed chemical agents, so no agent surface concentration measurement standards are presented in Table 2-2. However, a recent report from the Centers for Disease Control and Prevention (CDC), which reviewed the chemical agent monitoring programs at the final four CMA demilitarization facilities, specifically recommended that CMA develop a health-based agent hazard level (in milligrams per square meter) for anticipated surface contamination measurements using surface wipe samples on potentially contaminated structural surfaces (CDC, 2010). If such a health hazard standard were to be developed and accepted by relevant regulators, it could be used in conjunction with appropriately calibrated direct surface composition instrumental measurements to determine if secondary waste and structural materials are clean enough to be released for disposal. The CDC suggested that wipe sample evaluations could be used to assess agent contamination on nonporous materials. As discussed in subsequent sections, the newer direct composition measurement technologies may also be able to assess contamination levels of porous materials such as activated charcoal. However, the preparation of accurate agent-contaminated surface calibration standards using appropriate substrates (e.g., concrete, carbon, etc.) is a very challenging task. In the absence of such standards, direct surface contamination measurements will still produce semiquantitative relative distributions of surface contamination, which may nonetheless be valuable guides for material screening and decontamination activities.

**Finding 2-2.** No CMA or ACWA standards have been established for surface contamination similar to the airborne agent concentration exposure limits, from which vapor screening levels have been adopted. If accepted by the CDC and relevant state regulators, a health-based agent-contaminated surface hazard level measured in mass per unit area by a new, direct surface contamination measurement technology and suitable agent-contaminated surface calibration standards could be useful in clearing secondary waste materials during ACWA disposal operations and/or structural materials during closure. However, reliable agent-contaminated surface calibration standards may be difficult to produce.

In the ACWA facilities, all areas through which munitions pass and where agent could be present at some time will be monitored for airborne agent in NRT using MINICAMS, and archival data will be obtained with Depot Area Air Monitoring System (DAAMS) thermal desorption tubes, which are collected for later gas chromatography/mass spectrometry (GC/MS) analysis in the laboratory. These agent monitoring methods were described in more detail in the NRC report *Monitoring at Chemical Agent Disposal Facilities* (NRC, 2005a).

Work areas in a chemical agent destruction facility are classified by their likely contamination level using the categories listed in Table 2-4. Worker protection requirements may vary for a given contamination level, depending on the presence of liquid agent and the tasks at hand. The requirements for worker protective clothing and equipment are detailed in Chapter 4 of U.S. Army (2008) and U.S. Army (2011a).

Rooms and corridors adjacent to a Category A or B area are usually classified as Category C. In Category C or D areas, workers must have a protective mask readily available in the event of an accidental agent leak. All workers in Category A areas wear the minimum level of protection required. If liquid agent can be present, then Level A demilitarization protective ensemble (DPE) suits must be worn. The workers are sealed into plastic suits made of 20-30 mil polyvinyl chloride (PVC) that completely enclose them. They don the suits through a slit in the back after which the opening is hermetically sealed by heat. Inside the suit, they also carry on their back a 10-15 minute supply of air to be used in an emergency egress. At waist level, there is a fitting on the DPE suit to attach an umbilical air hose. Workers operate in teams of two in Category A areas. The team is taxied from the DPE donning area to the point where an entry will be made. A third worker in Level B protective ensemble remains outside in the entry vestibule in case of an emergency. As the two workers navigate through the Category A area they plug the end of their air hoses into the nearest air supply valve, one of many that are installed throughout the area. The workers can stay in the area for a maximum of 2 hr (U.S. Army, 2008). Before exiting the area they spray one another with decontamination solution to remove any gross liquid contamination from their DPE suits. Under some conditions, DPE suits are monitored for agent contamination using a MINICAMS detector with a sampling wand to determine if decontamination is complete. DPE suits are removed by cutting them open in the change vestibule. Contaminated suits may be returned to the Category A area for later treatment. These procedures may vary depending on worker condition, site operating procedures, and regulations.

TABLE 2-4 Room Contamination Requirements Using Near-Real-Time Monitoring of the Vapors in the Room

| Category | Definition | Typical Concentrations (AELs) | Typical Rooms with This Category | Required Protective Clothing |
|---|---|---|---|---|
| A | A toxic area supported by a cascade ventilation in system designated for expected chemical agent liquid and/or vapor under normal conditions | IDLH | Toxic areas with agent present and access to the atmosphere, such as munition breaching areas | OSHA/EPA Level A protection (typically DPE) |
| B | A toxic area supported by the cascade ventilation system designated for expected chemical agent vapor under normal situations but expected chemical agent liquid under off-normal situations. | <1 VSL | Explosive containment rooms (ECRs), transfer and holding areas | Typically OSHA/EPA Level A, B, or C protection, depending on work tasks |
| C | An area adjacent to Category A or B areas, that is supported by a cascade ventilation system designated for expected chemical agent vapor under off-normal situations only. Chemical agent liquid is not expected under normal situation or off-normal situation. | <STEL with confirmation monitoring <WPL with confirmation monitoring | Work and process support areas such as the unpack area and observation corridors | Typically OSHA/EPA Level D (with escape mask readily available) |
| D | An area where chemical agent liquid or vapor are not expected under normal or off-normal situations. | <STEL when single contained chemical warfare materiel and personnel present | Facility support areas such as administrative areas and maintenance buildings | No general PPE requirement except to have escape mask readily available |
| E | An area maintained at positive pressure and where chemical agent liquid and vapor are precluded from entering. | No agent presence | Positive pressure support areas such as the control room and medical facilities. | No general PPE requirement except to have escape mask readily available |

SOURCE: Adapted from U.S. Army, 2011a.

# PUEBLO CHEMICAL AGENT DESTRUCTION PILOT PLANT

The chemical munitions stored at the PCD are artillery projectiles and 4.2-in. mortar rounds, all of which only contain mustard, a blister agent, in one of two forms, HD or HT. HD is distilled mustard and HT is a mixture of HD and T, which is essentially a dimer of H that lowers the melting point of the H. The Pueblo stockpile components are detailed in Table 2-5. A simplified flowchart for the destruction process is shown in Figure 2-1. Figure 2-2 is a more detailed flowchart depicting process waste streams and their disposal. Figure 2-3 displays the site map for the facility.

Pallets containing the projectiles will be transported from the depot's storage igloos to the munitions storage magazine (MSM) at PCAPP. Because munitions can only be transported during daylight hours and in good weather, an accumulation of munitions in the MSM is necessary to ensure round-the-clock operation at PCAPP. The storage igloos will be monitored for agent vapor before the pallets are removed; agent vapor levels will also be monitored in the MSM. Munition pallets will again be monitored for agent vapor before being removed from a transport vehicle after arrival at the unpack area in the enhanced reconfiguration building (ERB). Projectiles containing bursters will be moved to the reconfiguration room (Category B level), where the bursters will be removed by a linear projectile and mortar disassembly (LPMD) machine without disturbing the burster well that seals in the chemical agent. Uncontaminated energetics will be sent offsite for processing. Munitions with agent leaks (leakers) will be overpacked and any agent spills will be documented and decontaminated.

In cases where the burster of a leaker or reject projectile cannot be removed robotically, the entire munition will be disposed of without disassembling using an explosive destruction technology (EDT), an ancillary processing system. There are several types of EDTs, but all involve the destruction of the chemical munitions and their contents by detonation within an enclosed chamber capable of containing the blast. The EDT completely destroys both agents and energetics, reducing them to water, carbon dioxide, and mineral salts. Using an EDT to process problem munitions, such as leakers and rejects, avoids interfering with the higher throughput achievable for processing normal munitions through the main PCAPP facility.[4]

The reconfigured projectiles will then be transported robotically in munition transfer carts through a long corridor to the agent processing building (APB). In the APB, the munitions, still containing the burster well that seals the agent within the agent cavity of the munition casing, will be moved on trays to the munitions washout system (MWS) in a Category A area. A robot will take a projectile from a tray and place it into one of several cavity access machines (CAMs) in an inverted position. In the CAM, a hydraulic arm will dislodge the burster well by ramming it up into the munition casing and exposing the agent. The agent will then be drained from the munition casing and the interior of the agent cavity will be washed using a high-pressure water wand on the hydraulic arm.

The chemical agent removed from the munition will then be transferred to the agent neutralization system (ANS) area, where it will be neutralized with hot water. Hydrolysate will not be transferred from the ANS until it has been analyzed and verified

---

[4]A leaker is a munition that has leaked agent. A reject is a munition that for any reason cannot be robotically disassembled. For more information on EDTs, see NRC, 2006; 2009a.

TABLE 2-5  Chemical Weapons Stockpile of HD- or HT-Filled Munitions at Pueblo Chemical Depot

| Munition Type | Chemical Fill (kg) | Energetics Content (kg) | Configuration |
|---|---|---|---|
| 105-mm cartridge, M60 | HD, 1.4 | Burster: tetrytol, 0.12<br>Fuze: M51A5<br>Propellant: M1 | Unreconfigured. Complete projectile includes fuze, burster. Propellant loaded with cartridge. Cartridges packed two per wooden box. |
| 105-mm cartridge, M60 | HD, 1.4 | Tetrytol, 0.12 | Reconfigured. Includes burster and nose plug, but no propellant or fuze. Repacked on pallets. |
| 155-mm projectile, M110 | HD, 5.3 | Tetrytol, 0.19 | Includes lifting plug and burster but no fuze. On pallets. |
| 155-mm projectile, M104 | HD, 5.3 | Tetrytol, 0.19 | Includes lifting plug and burster but no fuze. On pallets. |
| 4.2-in. mortar, M2A1 | HD, 2.7 | Tetryl, 0.064<br>Propellant: M8 | Includes propellant and ignition cartridge in a box. |
| 4.2-in. mortar, M2 | HT, 2.6 | Tetryl, 0.064<br>Propellant: M8 | Includes propellant and ignition cartridge in a box. |

NOTE: The M1 propellant present in 105-mm cartridges that have not been reconfigured (as defined in the column "Configuration") is present in M67 propelling charges—that is, granular propellant contained in bags as specified in MIL-DTL-60318C.

SOURCE: Adapted from NRC, 2008a.

that at least 99.9999 percent of the agent has been destroyed. The ANS area is classified as Category A and will be contaminated with agent. The area affected by accidental agent spills is minimized by the sumps constructed around the CAMs and any other vessel that contains agent or hydrolysate. Concrete surfaces have a polymer-based coating to help prevent the agent from being absorbed into them.

From this point on in the process, only residual amounts of agent exist in the hydrolysate waste stream. The hydrolysate produced from the neutralization of mustard agent contains thiodiglycol, a major hydrolysis product that must be destroyed to satisfy the Chemical Weapons Convention. The hydrolysate will be transferred to and treated in immobilized cell bioreactors, where bacteria will convert the hydrolysate organic compounds, including thiodiglycol, to water, carbon dioxide, and sludge.[5]

---

[5]In this context, sludge refers to precipitated solid matter composed of dead microorganisms, insoluble salts, and other low-solubility materials produced during the bioprocessing of mustard agent hydrolysate.

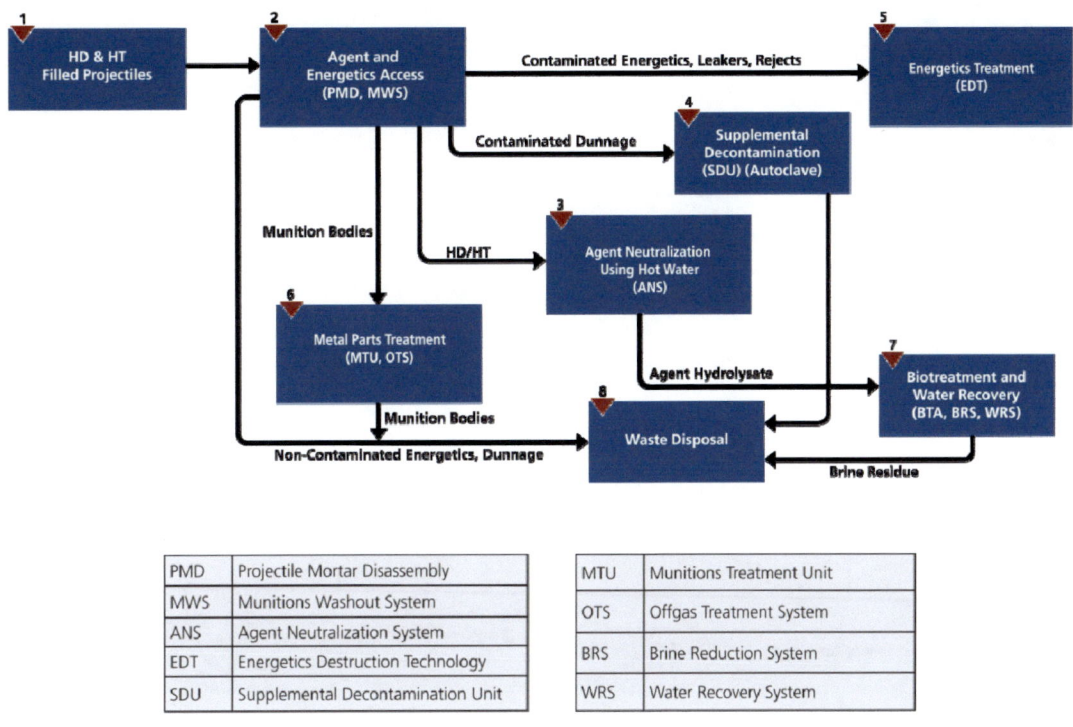

FIGURE 2-1 PCAPP munitions process flow chart. SOURCE: Sean Smith, USAE ACWA Systems Engineering, "PCAPP Overview," presentation to the committee on February 22, 2011. Note: The numbers above the boxes are simply sequential indicators that complement the arrows and act as guides for the multifaceted destruction progression from box to box.

After the agent and any residual solid heel are removed, the projectile bodies containing the burster wells are placed in other trays and moved to the munitions treatment unit (MTU), where they will be decontaminated by heating to at least 1000°F for over 15 min prior to being released. The MTU is a long, electrically heated muffle furnace with a conveyor that will slowly move projectile bodies from one end to the other as they are treated. The front end of the MTU is in a Category A area, while the back end is contained in a Category C area. The front end of the MTU will be agent contaminated.

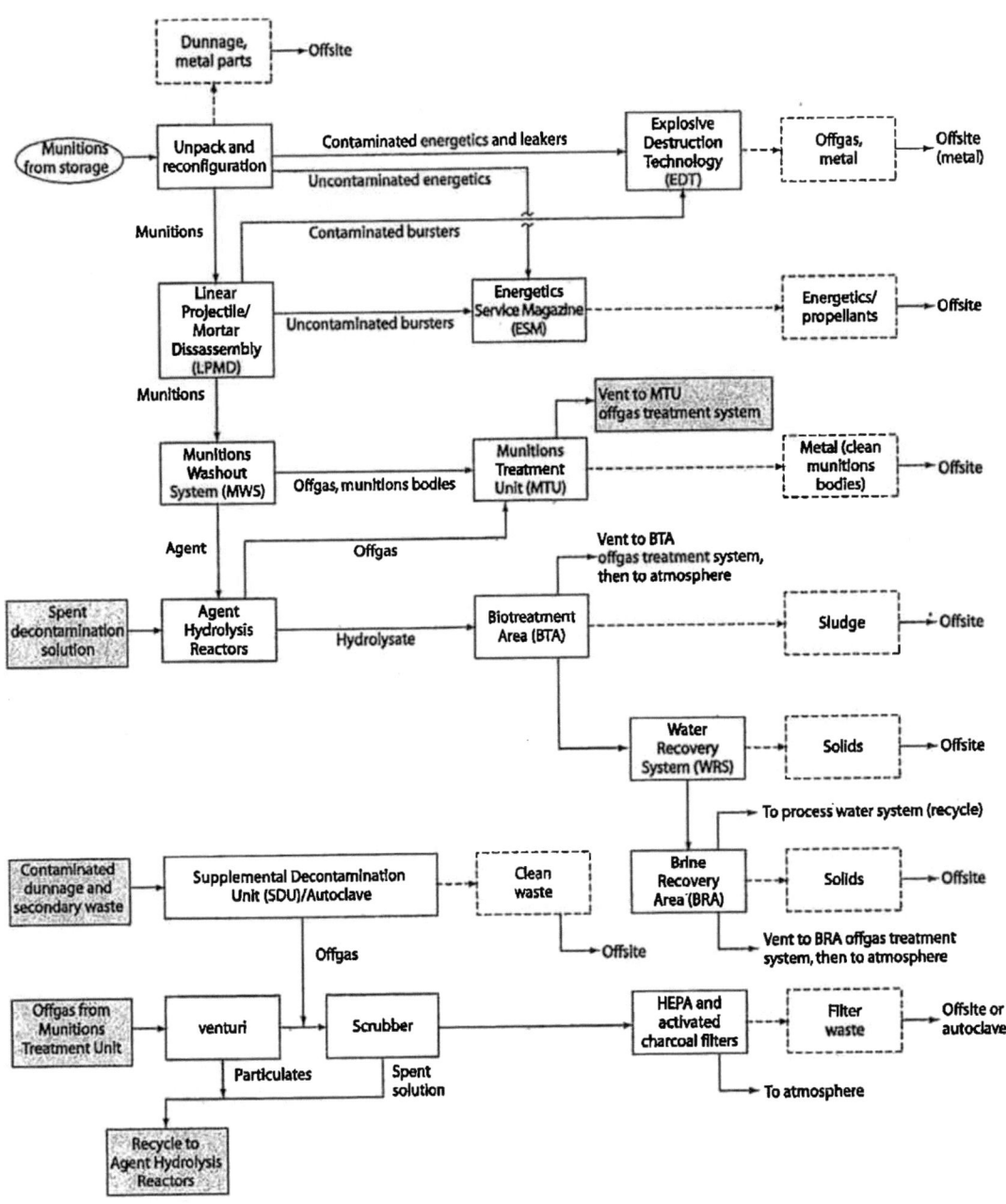

FIGURE 2-2 Process and waste stream diagram for PCAPP. SOURCE: NRC, 2008a.

FIGURE 2-3 PCAPP site layout. SOURCE: Sean Smith, USAE ACWA Systems Engineering, "PCAPP Overview," presentation to the committee on February 22, 2011.

# BLUE GRASS CHEMICAL AGENT DESTRUCTION PILOT PLANT

The chemical munitions stockpile stored at BGAD is more diverse than that at the PCD. It includes M55 rockets weaponized with GB or VX nerve agents, and several types of projectiles that contain mustard agent H or GB or VX nerve agents. Consequently, the process for destroying the munitions in the BGAD inventory is more complex than the one at PCAPP. The contents of the stockpile stored at the Blue Grass Army Depot are shown in Table 2-6, and the simplified flowchart for the destruction process is shown in Figure 2-4. Figure 2-5 presents a more detailed flowchart with the waste streams and disposal processes indicated. Figure 2-6 displays the site map for the facility.

The description of the demilitarization process can be separated into two sections: destruction of the projectiles, which is similar to the PCAPP design (and is for that reason not repeated here), and the treatment of the M55 rockets.

## M55 Rocket Processing

The M55 rockets filled with the volatile nerve agent GB will be the first munitions to be destroyed at BGCAPP because they pose the highest storage and processing risks. Each rocket contains about 19 lb of a double-base propellant (approximately 80 percent nitrocellulose and 20 percent nitroglycerine) and 10 lb of GB. The rockets will be transported from storage igloos into the unpack area, where personnel will remove them from the pallets. The rockets will be monitored for agent leaks in the igloos and again after transportation. (Usually, leaks are vaporized agent.) If agent vapor is detected outside the firing tube, the rocket is overpacked and returned to storage for later disposal. Leaker events are documented to inform subsequent decontamination actions.

The rockets, still contained in their fiberglass shipping and firing tubes (SFTs), will be placed on a conveyor and moved to the explosion containment vestibule (ECV) and onto a rocket cutting machine (RCM). First, the propellant motor section at the back end will be separated from the rocket warhead by cutting through the SFT and rocket body with a pipe-cutter-like device. The first cut is only deep enough to cut open the SFT so that it can be removed. Then, the cut will be deepened enough to breach the outer body of the rocket, allowing the warhead and motor sections to be separated. Uncontaminated propellant sections and the two firing tube sections will be shipped off-site for disposal. Up to this point, no agent contamination should exist unless an unexpected event occurs and the warhead is breached. Thus, the ECV is considered a Category B area.

After separation, the rocket warheads will be transferred to a rocket shearing machine in the explosive containment room (ECR). The top and bottom of the rocket will be punctured and the agent is drained. The warhead cavity will then be washed out with a high-pressure (430 psig) water system to remove residual agent as well as any gelled or crystallized material that may have formed during storage. The drained warhead will then be sheared into four segments in the rocket shear machine.

TABLE 2-6 Description of the Chemical Weapons in the BGAD Stockpile

| Munition Type | Chemical Fill (lb) | Energetics Content (lb) |
|---|---|---|
| 155-mm projectile, M110 | H, 11.7 | Tetrytol, 0.41 |
| 8-in. projectile, M426 | GB, 14.4 | None |
| 115-mm rocket, M55 | GB, 10.7 | Composition B, 3.2; M28 propellant, 19.1 |
| 115-mm rocket warhead, M56 | GB, 10.7 | Composition B, 3.2 |
| 155-mm projectile, M121/A1 | VX, 6 | None |
| 115-mm rocket, M55 | VX, 10.1 | Composition B, 3.2; M28 propellant, 19.1 |
| 115-mm rocket warhead, M56 | VX, 10.1 | Composition B, 3.2 |

SOURCE: Adapted from NRC, 2008a.

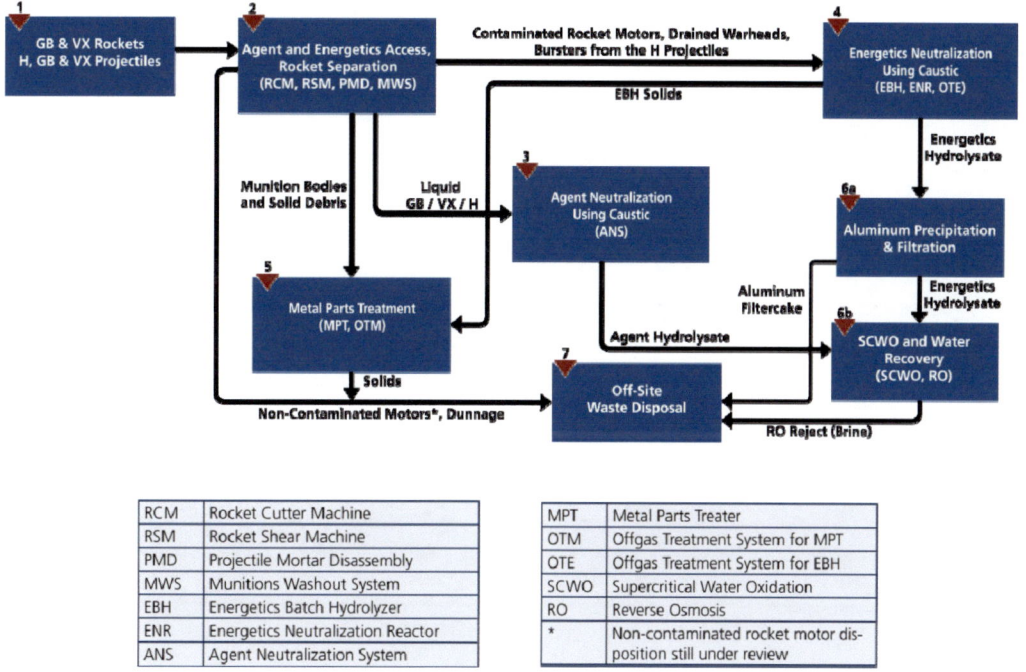

FIGURE 2-4 BGCAPP munitions process flow chart. SOURCE: Darren Dalton, BGCAPP Systems Engineer, ACWA, "BGCAPP Site Project Overview," presentation to the committee on February 22, 2011. NOTE: The numbers above the boxes are simply sequential indicators that complement the arrows and act as guides for the multifaceted destruction progression from box to box.

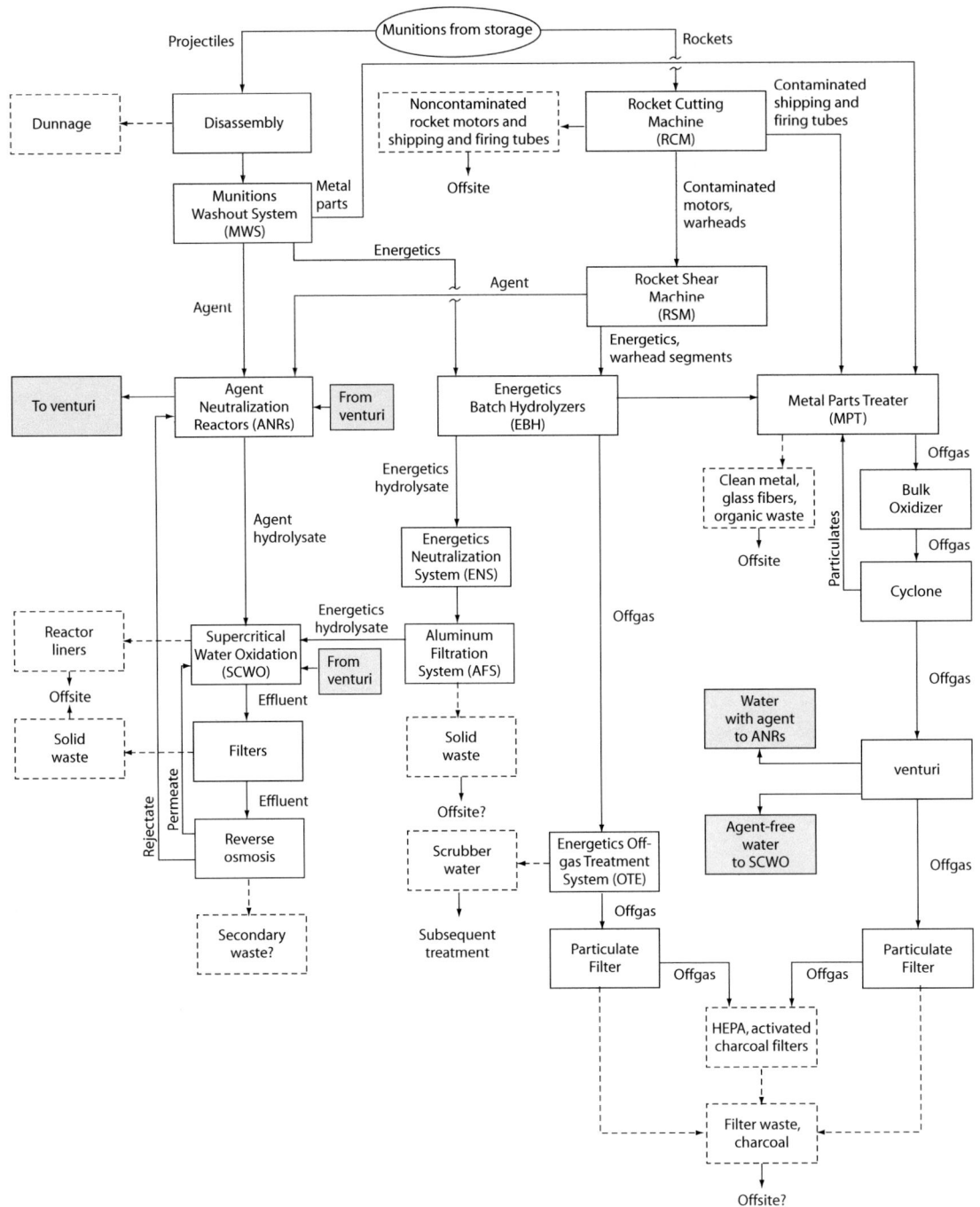

FIGURE 2-5 Process and waste stream diagram for BGCAPP. SOURCE: NRC, 2008a.

FIGURE 2-6 BGCAPP site layout. SOURCE: Darren Dalton, BGCAPP Systems Engineer, ACWA, "BGCAPP Site Project Overview," presentation to the committee on February 22, 2011.

After the reverse assembly (i.e., disassembly) is completed, two process streams are produced (see Figures 2-4 and 2-5):

- Agent for processing in the agent neutralization reactors (ANRs)
- Energetics for treatment in the energetic batch hydrolyzers (EBHs)

Chemical agent drained from the warhead will be transferred into holding tanks associated with the agent collection system. Water from rinsing the warhead will be sent to the holding tank for spent decontamination solution. From the holding tank, the chemical agent and rinse water are sent to an ANR, where the chemical agent will be neutralized. The hydrolysate from the ANR will then be sent for treatment to one of three supercritical water oxidation (SCWO) units, where its organic constituents will be oxidized to water, carbon dioxide, and salts.

After shearing, each rocket segment will be dropped into a bucket. Sheared segments will include the burster and the fuze. The buckets will be conveyed to the EBH room. The three EBHs are large rotating vessels, similar to the drum of a cement mixer truck, that have discontinuous helical flights used to mix the components as the EBH rotates.. Once in the EBH room, a robot will pick up each bucket and raise it to a platform near the top of the EBHs. A second robot will then move the bucket from the platform to an EBH, where the contents are dumped into the EBH. This area will be Category A, susceptible to both liquid and vapor contamination.

Prior to the addition of metal parts and energetics, the EBHs are partially filled with water followed by 50 percent caustic, to produce a concentration of 39.5 percent caustic, then heated to 240°F. After processing the metal parts and energetics for the specified time, the direction of rotation of the EBH drum will be reversed, lifting the metal parts out of the EBH and dropping them onto the vibrating screen belt of a horizontal conveyor. Any liquid will pass through the screen and be collected. When this operation is completed, the rotation speed of the vessel will be increased, allowing the liquid to be removed from the EBH through a wire screen that catches any remaining solids.

The metal parts from the EBH are sent to the metal parts treater (MPT), where they will be decontaminated by heating them with superheated steam to over 1000°F for more than 15 min. After treatment, the metal parts can then be sent off-site for recycling or to a landfill. The hydrolysate from the EBHs will be sent to one of the three energetics neutralization reactors (ENRs), where the neutralization is completed. The contents will remain in the ENR until it has been verified that all energetics and any agent present have been neutralized. The hydrolysate from the EBHs will then be sent to the SCWO units, where the products are oxidized to water, carbon dioxide, and salts. The energetics hydrolysate will be blended with the agent hydrolysate to prevent salt components from potentially solidifying within the SCWO unit; the blending forms a eutectic that lowers the melting point of the blended components below the temperature of the SCWO reactors. The inner liner of the SCWO reactors is made of titanium. For processing the GB blended feed, this liner is scheduled for replacement every 5 weeks, and for the VX blended feed, it will be replaced every 2.6 weeks. For the H blended feed, the liner is not expected to require replacement since the expected replacement interval of 75 weeks for this feed exceeds the length of time needed for processing the amount of H mustard hydrolysate that would be generated at BGCAPP.

## Projectile Processing

BGCAPP projectile processing will be similar to the processes used at PCAPP. The projectiles will be unpacked manually and conveyed into an ECR, where the bursters will be removed by the two LPMD machines and, if not contaminated, sent to an EBH.[6] The munition bodies, still containing their burster wells, will then be moved to the MWS. The projectile bodies will be processed in an MPT instead of an MTU.[7] The process from this point on will be identical to the process at PCAPP, described above, except that the hydrolysates will be sent to the SCWO units instead of being sent to bioreactors for treatment.

## DESCRIPTION OF THE HVAC SYSTEMS USED AT BOTH FACILITIES

The committee's statement of task specifically identifies activated carbon as a major secondary waste for which surface monitoring and analysis for agent contamination should be considered. This section briefly describes the uses of activated carbon at BGCAPP and PCAPP, which, in general, are similar to its use at other chemical agent disposal facilities.

The bulk of the activated carbon to be used at BGCAPP and PCAPP resides in the filter banks of the plants' cascaded HVAC systems. The HVAC system at each site is carefully designed to protect workers, the public, and the environment. Ambient air is circulated through the plant in a cascading manner from areas with no agent contamination to those with low contamination areas and finally to Category B and A areas, where the most contaminated plant air is expected. These latter areas are maintained at the lowest ambient air pressures to prevent backflow to less contaminated areas. More specifically, in the MDB, cascading air passes first through the Category C areas, including the observation corridors, then through the vestibules and airlocks and on to Category B areas, such as the ECR room containing the LPMD, and/or Category A areas, such as the MWS. The ambient air will be constantly monitored using MINICAMS and DAAMS tubes as it proceeds through areas of increasingly reduced atmospheric pressure. Finally, the air moves on to the carbon filter farm and then to the exhaust stack, leaving the plant. The design of the HVAC system for Category E areas such as the process control room, the medical facility, and so on, also accommodates carbon filters

---

[6]If the projectile is a leaker or a reject whose burster cannot be removed by the LPMD machines, it will be overpacked for later disposal processing, possibly by an EDT.

[7]The MPT is unique FOAK machinery designed for particular application at BGCAPP; the MTU to be used at PCAPP is an adaptation of a metal annealing oven or, as previously indicated, a continuous-belt muffle-type oven with material-handling capabilities at the feed and discharge ends. Additional information concerning the MPT (and the MTU) can be found in NRC (2008b).

installed on the air intakes while providing for the maintenance of positive air pressure within these areas.

The filter farms are identical to those used at the baseline incineration facilities. The description given here is based on a previous NRC report, *Disposal of Activated Carbon from Chemical Agent Disposal Facilities* (NRC, 2009b). An aerial view of a typical filter farm is depicted in Figure 2-7. A typical farm has eight or nine filter units. Since the operating schedules for PCAPP and BGCAPP are relatively short, it is expected that the carbon will not have to be changed out before closure. Also, at BGCAPP only two units may be used at any one time. Two other units will be used after the first agent disposal campaign is completed. Figure 2-8 shows a side view of a vestibule with the doors leading into the spaces between the individual filter banks. Figure 2-9 is a schematic representation of airflow through a filter unit. The airflow through the unit takes the following path:

1. Through a coarse 85-micron particulate filter to remove large particulates.
2. Through a HEPA filter to remove any remaining particulates.
3. Through six banks of carbon filters with air spaces between them.
4. Through a second HEPA filter to catch any carbon particulates escaping from the carbon filters.
5. Through the large induced-draft fan that drives the whole HVAC system by sucking the air out.
6. Up the exhaust stack and into the atmosphere.

Each filter bank comprises 48 trays weighing about 100 lb each and containing about 48 lb activated carbon. A tray is depicted in Figure 2-10 and the airflow through the tray in Figure 2-11.

MINICAMS monitor the air between banks 1 and 2, 2 and 3, 3 and 4, and 4 and 5, as shown in Figure 2-9. Some variations on these procedures may occur after negotiations with the state regulators. If agent vapor at one STEL (see Table 2-1) were detected between banks 2 and 3, the carbon in banks 1 and 2 would be immediately replaced. Banks 3 through 6 are never likely to be contaminated and can probably be disposed of based on generator knowledge (see Box 2-1), but banks 1 and 2 are likely to be heavily contaminated before breakthrough. It is not expected that the carbon will ever have to be changed during the relatively short operations schedule. The carbon may have to be treated as contaminated waste and be disposed according to the each state's specific regulations.

The adsorbed agents on the carbon can be analyzed by solvent extraction (EPA, 2007a), followed by GC/MS quantification (EPA, 2007b). Studies indicate that both the nerve agents, GB and VX, and the mustard agent adsorbed on carbon degrade over time at varying rates and through different catalytic pathways involving water vapor, which is also adsorbed onto the carbon (NRC, 2009b). However, without direct analysis, the carbon must be treated as contaminated. A rapid method of scanning the surface to indicate the presence or absence of agent would greatly simplify the disposal process. But its use would have to be approved by state regulators.

FIGURE 2-7 The nine activated carbon filter units for the MDB HVAC system. SOURCE: NRC, 2009b.

FIGURE 2-8 Vestibule on the side of an MDB HVAC unit. SOURCE: NRC, 2009b.

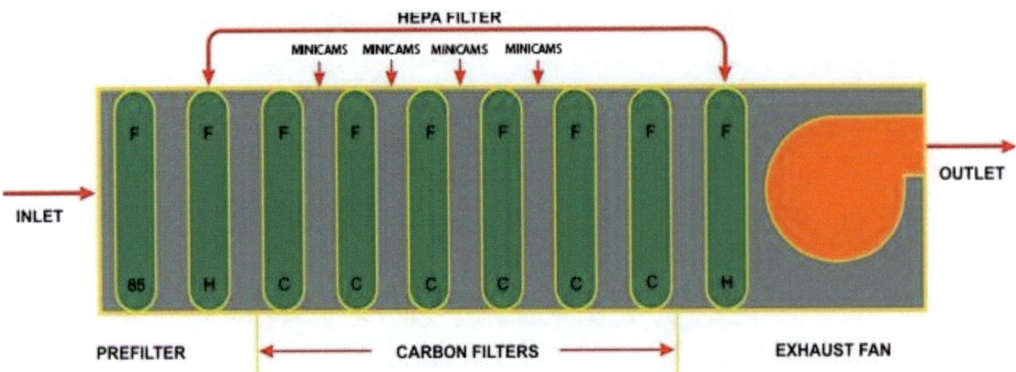

FIGURE 2-9 Schematic representation of airflow through the six filter banks that make up each MDB filter unit. Carbon filters each contain 48 filter trays arrayed in six columns and eight rows with each tray oriented in horizontal position. The 85 indicates 85 percent efficiency for the particulate prefilter; H indicates HEPA filter; F indicates filter; and C indicates carbon filter. SOURCE: Adapted from NRC, 2009b.

FIGURE 2-10 A filter tray. SOURCE: NRC, 2009b.

Other uses of carbon at BGCAPP and PCAPP involve considerably smaller amounts; however, these uses produce a wide range of contamination levels. For example, activated carbon is used in the canisters in workers' protective masks. These may or may not become contaminated depending on where and how they are used. Activated carbon is also used in the filters on the venting outlets of the agent collection system (ACS), where agent drained from munitions is collected before neutralization. The ACS carbon is expected to become highly contaminated.

Current methods used to assess agent contamination of machinery, equipment, and waste streams at U.S. Army chemical weapons storage and demilitarization facilities are discussed in more detail in the next chapter.

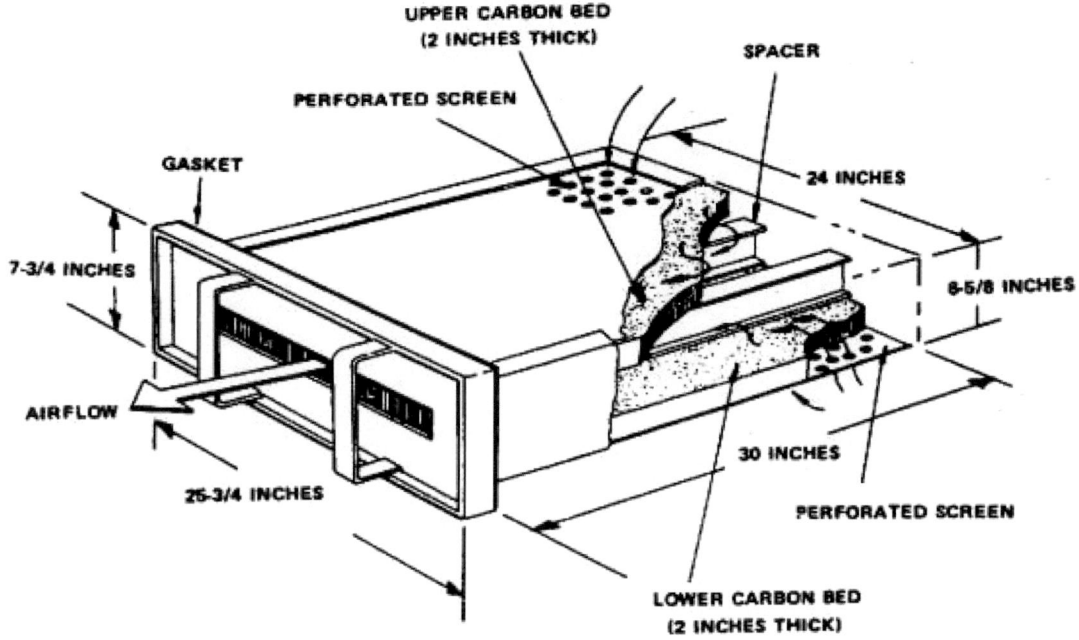

FIGURE 2-11 Airflow path through a filter tray. SOURCE: NRC, 2009b.

---

**BOX 2-1**

**DEFINITION OF GENERATOR KNOWLEDGE**

According to ACWA's Monitoring Concept Plan document, generator knowledge is defined as follows (U.S. Army, 2011a, p. 56):

Knowledge of the item's operational and contamination history shall be used to assess and bound the possible agent contamination of a solid or liquid material. The assessment may be based on factors such as location of the item and its proximity to agent in the facility, use in the facility's operational processes, and worker knowledge of the item's history. Generator knowledge shall also use supporting records and documents to provide relevant historical details for definitive evidence of contamination assessment. Examples of records for contamination categorization include operational, maintenance, and/or monitoring documents that describe the system associated with the item supported (such as airlock and room agent readings, LSS [life support system] air readings) and any other records that would indicate the potential agent contamination of the item or material.

# 3
# Agent Monitoring Practices for Waste Generated at BGCAPP and PCAPP

In this chapter, the committee briefly describes plans for the overall monitoring of waste streams expected to be generated at the Blue Grass Chemical Agent Destruction Pilot Plant (BGCAPP) and the Pueblo Chemical Agent Destruction Pilot Plant (PCAPP) and uses this as a basis for identifying opportunities to employ new ambient ionization surface-sensitive mass spectrometric techniques that offer potential benefits at these two Assembled Chemical Weapons Alternatives (ACWA) pilot plants. The committee recognizes that plans for operating these facilities are well under way and that any implementation of new technology may require time for adaptation and procedural vetting, etc. Given these considerations, surface measurement technologies may most likely be utilized during the later stages of operations, particularly during facility closure.

The committee wishes to point out that in examining the applicability of the new technologies discussed in this report, it is not calling into question the current monitoring protocols planned for these facilities. These have proven to be safe and effective during nearly two decades of chemical agent disposal campaigns at other chemical demilitarization facilities. Rather, the question is whether the new techniques could offer capabilities that beneficially augment the present techniques and perhaps lead to a reduction in the duration of the overall schedule.

This chapter is intended to identify monitoring opportunities and challenges. Chapters 4 and 5 address the suitability of the new ambient ionization mass spectrometric techniques to address those challenges. Several potential scenarios are presented in this chapter, where new measurement capabilities might facilitate faster, safer, or more efficient agent disposal and closure operations.

## WASTE ANALYSIS OVERVIEW

Chapter 2 presented an overview of the process flow plans for PCAPP and BGCAPP. This chapter addresses the plans for monitoring and analysis of those processes and the waste streams they will produce. This is not a comprehensive review because this subject was previously examined in the National Research Council (NRC) report *Review of Secondary Waste Planning for the Blue Grass and Pueblo Chemical Agent Destruction Pilot Plants* (2008a). Instead, the following sections are intended to provide a perspective on where additional monitoring tools might be useful. The ACWA pilot plants are full-scale facilities that face all of the normal state permitting requirements that any industrial facility must comply with in terms of dealing with effluents and waste streams. Since these general requirements are reasonably standard, this report focuses on the unique aspects of monitoring for chemical agent contamination.

## WASTE GENERATION AND MONITORING OVERVIEW

The various waste streams generated at BGCAPP and PCAPP can be classified as primary or secondary:

- *Primary waste streams*. Those encountered in conducting primary operations for disposal (e.g., explosives removed and agent drained from munitions) that are treated on-site, and
- *Secondary waste streams*. Those generated by activities either in support of or downstream of the primary processes for agent and energetics destruction—for example, activated carbon, used demilitarization protective ensemble (DPE) suits, dunnage, and so on that ultimately leave the facility. These wastes include additional materials produced during facility closure such as demolished concrete.

The characterization of the expected secondary wastes from the Pueblo and Blue Grass facilities and the planning for their disposal were described in detail in NRC, 2008a. The purpose of that study was to "provide PMACWA with a technical appraisal of its evolving plans to safely and efficiently handle, treat, and ultimately dispose of the waste materials that remain following the destruction of the assembled chemical weapons stored at PCD and BGAD" (p. 7). In the course of preparing the present report, the ACWA Monitoring Committee received additional information on this subject.[1]

A general overview of the waste monitoring plan for PCAPP is shown in Figure 3-1, which itemizes the planned processes and corresponding analytical monitoring and sampling methods. A corresponding diagram for BGCAPP was unavailable to the committee, but the committee believes the PCAPP plan is reasonably illustrative and representative for the purpose of this report. The analytical methods are categorized by the primary process from which they originate and further identified by the type of sample and key measurement parameter.

In the case of generator knowledge, sampling is not needed if there was no opportunity for agent contamination to have occurred based on the Army's criteria shown in Box 2-1. In other cases, headspace samples are taken after the waste is bagged or tented and allowed to equilibrate. The sample is then taken from the vapor space above the waste within the tented area, as described in the next section. In most cases, the methods are derived from standard EPA methods; however, some methods were

---

[1] This includes a presentation to the committee by Gary Groenewold, a member of the Committee to Review Secondary Waste Disposal and Regulatory Requirements for the Assembled Chemical Weapons Alternatives Program, "Sources and Amounts of Agent-Contaminated Wastes," on February 23, 2011, and information provided by ACWA staff in response to committee questions or communications during site visits.

developed specifically for the chemical agent destruction program. These methods are indicated by the inclusion of the PCAPP method number (e.g., PCAPP-204).

The secondary waste streams anticipated during normal operations for PCAPP, which only has munitions containing mustard agent, are summarized in Table 3-1. Table 3-2 shows the anticipated secondary waste expected from closure of PCAPP. The corresponding secondary wastes expected from BGCAPP are presented in Tables 3-3, 3-4, and 3-5. The waste streams are divided into "agent contaminated" (>1 VSL) and "clean" (<1 VSL) based on the airborne exposure limits (AELs) and vapor screening levels (VSLs) detailed in Tables 2-1 and 2-2. The distinction between "agent-contaminated" and "clean" is important because it determines the type of handling required and the type of waste treatment, storage, and disposal facility (TSDF) to which the material can be sent.

At the time this report was prepared, PCAPP was further along on construction than BGCAPP and had an approved waste analysis plan (WAP) in place.[2] A draft WAP submitted by BGCAPP as part of its Resource Conservation and Recovery Act (RCRA) permit application for BGCAPP had not been approved at the time of this writing. From a monitoring perspective, the only major difference between the two WAPs anticipated by the committee is that at BGCAPP, a sequence of agents needs to be monitored as the different agents are processed. The guiding principle for handling waste that may be agent-contaminated was summarized on page 34 in NRC, 2008a:

> Under the WAP filed with the Colorado Department of Public Health and Environment (CDHPE), PCAPP will use generator knowledge as the primary means of characterization, with direct sampling and analysis used to verify process knowledge. Agent monitoring is conducted in accordance with the Army's AEL guidance dated June 18, 2004). There are three approaches for classifying and disposing of secondary waste relative to its contamination by agent:
>
> 1. The waste is containerized and its headspace is monitored to determine the appropriate classification: or
> 2. The waste is assumed to be agent-contaminated and is decontaminated in accordance with the RCRA permit or regulations; adequate decontamination (<1.0 VSL) is verified via monitoring at the SDU [supplemental decontamination unit] or autoclave, whereupon it is reclassified as "clean and shipped offsite; or
> 3. The waste is assumed to be agent-contaminated and is shipped offsite to a facility permitted to receive such wastes.

For the purposes of this report, the committee again notes that the Army's currently accepted means of characterizing agent-contaminated waste are generator knowledge or the AELs established by the Army as described in Chapter 2. The AELs are the basis for (1) quantifying the agent contamination measurements known as VSLs for wastes subjected to controlled air monitoring and for (2) the established waste control limits (WCLs) for wastes that must be analyzed by extractive procedures.

---

[2]For example, see http://www.cdhpe.state.co.us/hm/pcd/adminrecord.htm, accessed on September 12, 2011.

TABLE 3-1 Projected Amounts of Mustard-Agent-Contaminated Secondary Waste from Normal Operations at PCAPP

| Stream Description | Amount (lb) | |
| --- | --- | --- |
| | <1 VSL | >1 VSL |
| Wood | 0 | 56,906 |
| Fiber tubes, additional packing material, metal strapping, miscellaneous metal | 0 | 0 |
| TAP gear | 9,639 | 6,709 |
| Steel | 0 | 0 |
| Lead alloy | 0 | 0 |
| Aluminum | 18 | 53 |
| Brine reduction | 0 | 0 |
| Water recovery thickener residue | 0 | 0 |
| Energetics | 0 | 0 |
| Brass and copper wire | 0 | 0 |
| Charcoal from PPE mask containers | 0 | 2,583 |
| Inert bulk solid waste | 15,421 | 35,790 |
| Halogenated waste | 3,153 | 2,661 |
| DPE suits | 121,514 | 81,010 |
| Waste oils and spent hydraulic fluid | 2,416 | 400 |
| Leather | 437 | 197 |
| Absorbents | 1,534 | 3,554 |
| Paper/fiberglass/rubber | 0 | 0 |
| Polystyrene and polyethylene | 669 | 2,318 |
| Combustible solid waste | 2,827 | 2,382 |
| Waste paint sludge | 915 | 455 |
| Dry cell batteries | 1,828 | 203 |
| Lead acid batteries | 1,219 | 135 |
| Mercury-containing lighting | 259 | 29 |
| Total | 161,849 | 195,385 |

SOURCE: PCAPP answers to Question Set 5 posed by the ACWA Secondary Waste Committee, March 11, 2008.

TABLE 3-2 Projected Amounts of Mustard-Agent-Contaminated Secondary Waste from Closure at PCAPP

| Stream Description | Amount (lb) <1 VSL | Amount (lb) >1 VSL |
|---|---|---|
| Wood | 0 | 0 |
| TAP gear | 3,704 | 412 |
| Steel | 0 | 0 |
| Aluminum | 21 | 7 |
| Brine reduction | 0 | 0 |
| Water recovery thickener residue | 0 | 0 |
| Propellant | 0 | 0 |
| Brass and copper wire | 0 | 0 |
| Charcoal | 0 | 1,000 |
| Inert bulk solid waste | 262,351 | 259,498 |
| Halogenated waste | 27,946 | 25,910 |
| DPE suits closure | 47,050 | 31,366 |
| Waste oils and spent hydraulic fluid | 927 | 164 |
| Leather | 147 | 98 |
| Absorbents | 350 | 3,153 |
| Paper/fiberglass/rubber | 0 | 0 |
| Polystyrene and polyethylene (poly drums and 5-mil poly bags) | 0 | 785 |
| HEPA/prefilters | 9,500 | 28,500 |
| HVAC | | |
|     Filtration charcoal | 30,690 | 3,410 |
|     Filter plenums | 15,300 | 1,700 |
|     Filter ductwork | 9,000 | 1,000 |
| Concrete | 38,775 | 12,925 |
| Combustible solid waste | 26,359 | 26,503 |
| Waste paint sludges and other sludges | 0 | 531 |
| Dry cell batteries | 707 | 79 |
| Lead acid batteries | 472 | 52 |
| Mercury-containing lighting | 100 | 11 |
| Total | 473,399 | 397,102 |

SOURCE: PCAPP answers to Question Set 5 posed by the ACWA Secondary Waste Committee, March 11, 2008.

TABLE 3-3 Projected Secondary Waste Streams for >1 VSL Agent-Contaminated Waste During Operations and Closure at BGCAPP

| >1 VSL Waste | Projected Totals (lb)[a] | |
|---|---|---|
| | Operations[b] | Closure[c] |
| Combustible solids | 5,242 | 30,879 |
| Metal | 24,737 | 449,457 |
| TAP gear/rubber | 555 | 390 |
| Halogenated plastic | 9,957 | 73,505 |
| Nonhalogenated plastic | 2,209 | 18,786 |
| Pre-HEPA filters | 1,044 | 13,140 |
| Agent collection system/spent decontamination solution sludge | 1,082 | 759 |
| Concrete | 0 | 50,053 |
| Foam wall panel | 0 | 31,371 |
| Special coatings | 0 | 4,052 |
| Aluminum | 0 | 2,149 |
| Overpack waste | 31,200 | 0 |
| Total | 76,068 | 674,540 |

NOTE: TAP, toxic agent protective; HEPA, high-efficiency particulate air; ACS, agent collection system; and SDS, spent decontamination solution.

[a] Totals calculated from estimated rate (lb/yr) data.
[b] BGCAPP operations are estimated to have a duration of 2.08 years.
[c] BGCAPP closure is estimated to have a duration of 1.46 years.

SOURCE: Adapted from PMACWA, 2006.

TABLE 3-4 Projected Secondary Waste Streams for <1 VSL Agent-Contaminated Waste During Operations and Closure at BGCAPP

| <1 VSL Waste (Unless Otherwise Noted) | Projected Totals (lb)[a] | |
|---|---|---|
| | Operations[b] | Closure[c] |
| Combustible solids | 2,623 | 22,014 |
| Metal | 22,087 | 571,717 |
| TAP gear/rubber | 1,066 | 1,066 |
| Halogenated plastic | 14,360 | 151,039 |
| Nonhalogenated plastic | 1,733 | 20,994 |
| 3X pre-HEPA filters | 82 | 5,084 |
| Sludge | 64 | 64 |
| 3X concrete | 0 | 79,993 |
| 3X foam wall panel | 0 | 50,136 |
| Special coatings | 0 | 6,475 |
| 3X aluminum | 48 | 3,435 |
| Total | 42,063 | 912,017 |

NOTE: 3X refers to a formerly used decontamination level that indicates that the item has been surface decontaminated by locally approved procedures, has been bagged or contained in an agent-tight container of sufficient volume to permit an air sample to be withdrawn while minimizing dilution with incoming air, and/or appropriate tests/monitoring have verified that concentrations are not above $0.0001$ mg/m$^3$ for agent GB, $0.00001$ mg/m$^3$ for agent VX, or $0.003$ mg/m$^3$ for H. Monitoring is not required for completely decontaminated and disassembled parts that are shaped simply (no crevices, threads, or the like) and are made of essentially impervious materials (such as simple lab glassware and steel gears) (NRC, 2007).

[a] Totals calculated from estimated rate (lb/yr) data
[b] BGCAPP operations are estimated to have a duration of 2.08 years.
[c] BGCAPP closure is estimated to have a duration of 1.46 years.

SOURCE: Adapted from PMACWA, 2006.

TABLE 3-5 Estimated Agent-Contaminated Waste Stream Summary for Operations and Closure at BGCAPP

| Waste Designation | Total Weight of the Waste (lb) |
|---|---:|
| Inert bulk solid waste | |
|     Metal | 1,243,545 |
|     Concrete | 152,369 |
| Aluminum waste | 6,685 |
| Foam core panels | 95,498 |
| Special coatings | 12,333 |
| Combustible bulk solid | |
|     Nonhalogenated plastics | 50,972 |
|     Tap gear | 4,555 |
|     HEPA filters and prefilters | 19,997 |
|     Adsorbents, cottons, rags, bulk | 4,477 |
|     Paper, wood, fiberglass, rubber | 63,794 |
| Halogenated plastics | 308,404 |
| Sludge | 1,997 |
| RCRA toxic metal-bearing waste | |
|     Paint chips | 121 |
|     Leather gloves | 224 |
|     Other | 1,000 |
| Waste oil and hydraulic fluids | 1,620 |
| Agent-contaminated activated carbon | 103,488 |
| Leaker campaign/overpack waste | 15,000 |
|     Total | 2,071,079 |

SOURCE: Adapted from BPBGT, 2006.

# MONITORING BASED ON VAPOR MEASUREMENTS

## Air Monitoring Instrumentation and Methods

The predominant time-honored methods for agent monitoring of both plant areas and wastes depend on measuring airborne concentrations. The instrumentation and methods used were most recently examined in the NRC report *Monitoring at Chemical Weapons Disposal Facilities* (2005a). Other NRC reports over the last two decades have also discussed monitoring instrumentation and methods in the context of those reports' subject matter.[3] The focus on disposal site monitoring and measuring airborne agent has been appropriate because vapor constitutes the most probable means of exposure for workers and is the only pathway by which the public could be exposed.

Miniature continuous air monitoring systems (MINICAMS) are the workhorse units chosen for airborne monitoring at PCAPP and BGCAPP; they provide near-real-time (NRT) data (5-15 min). Backup monitoring for confirmatory and historical purposes is provided by the depot area air monitoring system (DAAMS) collection tubes that adsorb and preconcentrate agent vapors from ambient air over a period of time. These collection tubes are transported to an on-site laboratory and their contents are regularly (daily or longer) flash-desorbed analyzed by gas chromatography/mass spectrometry (GC/MS). Results from the DAAMS tubes have a turnaround time of up to 72 hours and are thus not in real time or near real time, but do have the strength of providing a cumulative and historical record of agent vapor presence, even at very low concentrations.

In addition to their role in monitoring areas where agent contamination is expected, MINICAMS are also installed near to the areas where personnel might be exposed to agent. MINICAMS also are used for detection of leakers in storage igloos; in containers used to transport munitions from the igloos to the munitions storage magazine (MSM); and in transport vehicles from the MSM to the enhanced reconfiguration building. PCAPP anticipates that 132 MINICAMS will be required to support its operations.[4]

MINICAMS are also used for headspace monitoring of waste materials prior to shipment to an off-site TSDF, as mentioned above. The measurement cannot exceed the established WCLs, which are defined in terms of a VSL for each agent. As noted above, headspace monitoring of the waste involves placing it in an enclosure at a prescribed temperature for a sufficient amount of time to allow agent present on the solid to equilibrate with agent in the vapor. If the vapor-phase concentration is <1 VSL, the waste is deemed to be clean. Target release levels are generally somewhat lower than the WCLs to allow a margin for error and still be in compliance with the WCL values approved by state regulatory authorities.

One possible use of direct surface analysis using ambient ionization mass spectrometric technology could be to help resolve those instances where prior exposure

---

[3] Examples include NRC, 1994; 2002; 2005d; 2010.
[4] PCAPP Air Monitoring Strategies and Secondary Waste, discussion between Walter Waybright, PCAPP Laboratory Manager, and the committee, on June 28, 2011.

of materials to agent is in question. However, it should be noted that neither headspace concentrations (VSLs) nor concentrations determined from alternative extractive solvent analyses (discussed below) are directly and quantitatively derivable from surface concentrations. A first-order surface concentration measurement might entail obtaining a qualitative answer as to whether or not to analyze further. A conservative estimate of a corresponding gas-phase concentration could be made by assuming 100 percent desorption of the measured surface agent concentration multiplied by an estimated contaminated surface area to specify a total adsorbed agent mass, and then applying the ideal gas law to compute the potential gas phase agent concentration (mass per unit volume) in the headspace volume.

The immediate opportunities for ambient ionization monitoring techniques lie either in improving internal operation of the plant or in identifying materials with strongly absorbed agent that must be managed accordingly. By improving internal operations, the committee is referring to possible enhancements to worker safety and to potentially enabling the time needed for overall disposal and closure operations to be shortened.

## Extractive Analysis

Hazard analyses for work procedures conducted at chemical agent disposal facilities attempt to scrupulously avoid possibilities for dermal contact with chemical agent by workers. The use of appropriate-level personal protective equipment is among the prescribed means by which dermal contacts with agent are avoided. Consequently, the vapor-phase monitoring methods described above are aimed at controlling and providing valid measurements of the inhalation threat to workers posed by ambient airborne agent concentrations.

Meanwhile, certain waste materials (e.g., carbon and wood) are not amenable to accurately measuring the extent of agent contamination from headspace vapor monitoring owing to their physical properties (e.g., adsorptivity and absorptivity). The 2009 NRC report *Disposal of Activated Carbon from Chemical Agent Disposal Facilities* described the problem in using headspace vapor analysis for one such major waste stream (activated carbon) as follows (NRC, 2009b, p. 42):

> . . . to use this method to accurately measure agent loading on carbon requires measurement of the gas-phase concentration in equilibrium with the carbon and also requires knowledge of the adsorption isotherm for that agent under relevant conditions. In principle, if the adsorption isotherm is known, then the adsorbed-phase concentration or loading can be determined from the gas-phase concentration. Three issues associated with the use of headspace analysis must be considered to achieve a reliable analysis of agent loading on carbon. First, the gas-phase concentration of agent that would be in equilibrium with an agent loading of 20 ppb at ambient and even moderately elevated temperatures could be undetectable by head space analysis. Second, the adsorption isotherm would be needed to correlate loadings with gas-phase concentrations at agent loadings near 20 ppb. Third, a pure-component adsorption isotherm would not even apply

to the real system, which would contain co-adsorbed amounts of other components, such as water and degradation products.

Given the issues and problems surrounding accurate determinations of the degree of agent contamination for materials such as activated carbon using vapor-phase methods, the Army and various regulatory authorities have instead turned to requiring characterization of agent concentrations by analytical extraction procedures. These are described in EPA publication SW-846, *Test Methods for Evaluating Solid Waste, Physical/Chemical Methods*—specifically, Method 3571, "Extraction of Solid and Aqueous Samples for Chemical Agents," and Method 8271, "Assay of Chemical Agents in Solid and Aqueous Samples by Gas Chromatograph/Mass Spectrometry, Electron Impact (GC/MS/EI)" (EPA, 2007a,b).[5]

Several of the process monitoring methods summarized in Figure 3-1 involve extraction of a liquid sample aliquot containing potentially contaminated solid material, which would be subsequently analyzed in the laboratory by the techniques described in Methods 3571 and 8271 While they provide an alternative means to more accurately measure agent contamination, these methods involve time-consuming and difficult operations. That is, any method that requires taking an aliquot to a laboratory for analysis is far from real time, and it would seem desirable if it could be replaced with a real-time method. The new ambient ionization mass spectrometric methods offer the possibility of real-time agent contamination measurements of porous materials for the first time, giving the Army's chemical demilitarization community the opportunity to consider their utility. In addition, as shown in subsequent chapters, ambient ionization mass spectrometry methods are capable of quantifying low levels of relevant agents in liquid solutions. Thus, it may be feasible to analyze liquid extraction samples where they are collected without transport for laboratory analysis. For the purposes of this report, the committee's examination will focus on the most immediate and largest needs to characterize porous materials, activated carbon and concrete, which will be discussed in more detail in later sections of this chapter and in subsequent chapters.

## USE OF DPE SUITS DURING PLANT OPERATIONS

The largest single secondary waste category generated during normal operations at PCAPP (Table 3-1) is expected to be nonporous DPE suits: over 200,000 lb, of which 81,000 lb are anticipated to be >1 VSL. This represents roughly 40 percent of all anticipated waste with >1 VSL. Much of the DPE waste originates from routine maintenance entries into Category A areas, where agent is expected to be encountered. Another 78,000 lb or so of DPE waste are anticipated to be generated during closure operations at PCAPP. Unfortunately, DPE suits are not specifically itemized in Tables 3-3, 3-4, and 3-5 from BGCAPP, but the committee presumes that DPE suits, constructed

---

[5]SW-846 is the EPA Office of Solid Waste's (OSW's) official compendium of analytical and sampling methods that have been evaluated and approved for use in complying with the RCRA regulations. SW-846 functions primarily as a guidance document setting forth acceptable methods for the regulated and regulatory communities to use in responding to RCRA-related sampling and analysis requirements. Additional information is available at www.epa.gov/osw/hazard/testmethods/sw846/.

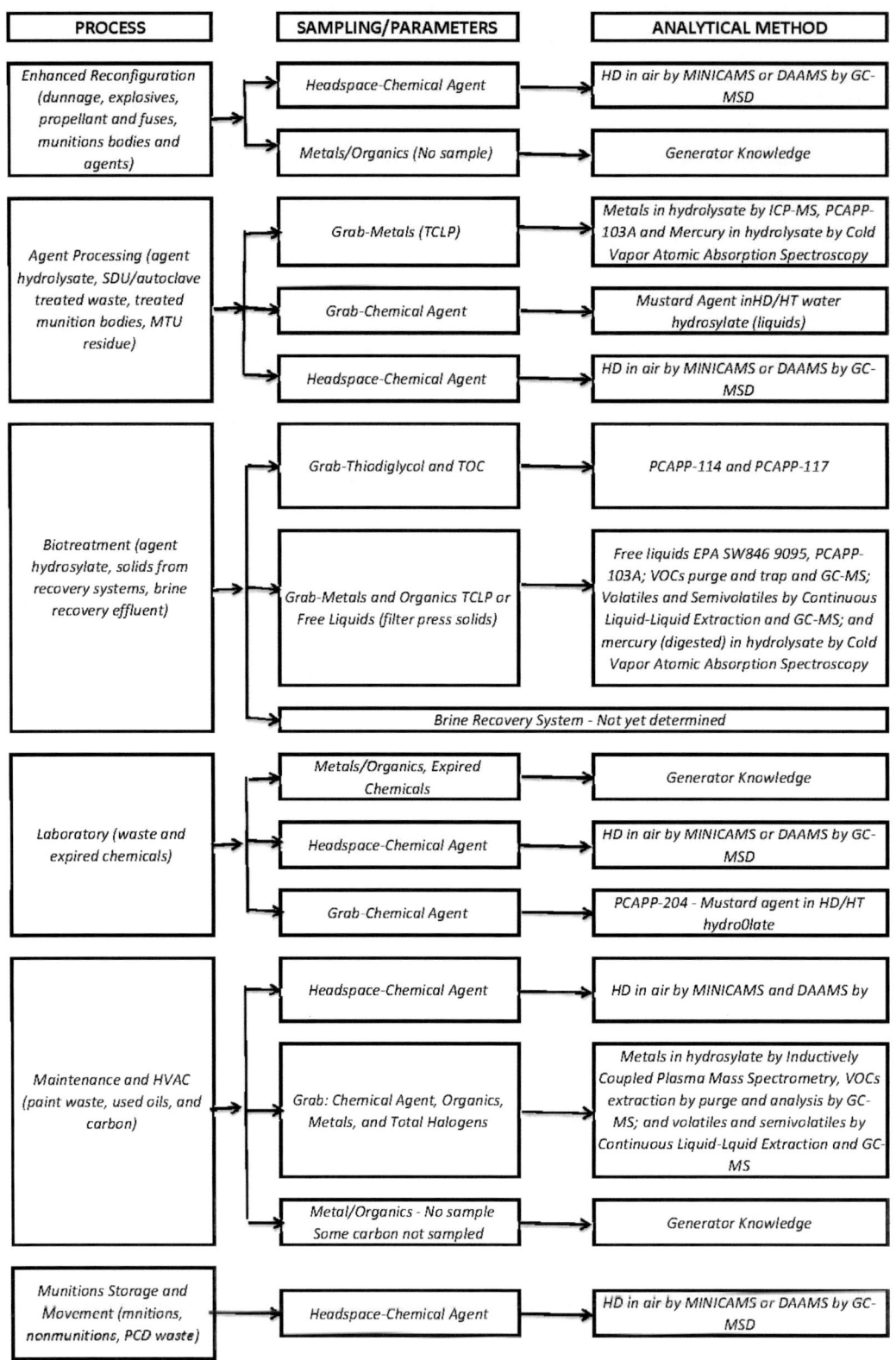

FIGURE 3-1 An overview of the analysis plan for PCAPP. SOURCE: Adapted from Walter Waybright, PCAPP, "PCAPP Air Monitoring Strategies and Secondary Waste Discussion," presentation to the committee on June 28, 2011.

from various thicknesses of PVC material, make up the bulk of the listed halogenated plastic waste.

**Finding 3-1.** The waste category designations used for tabulating waste streams at the Pueblo Chemical Agent Destruction Pilot Plant and the Blue Grass Chemical Agent Destruction Pilot Plant differ, thus making waste management comparisons between the two facilities difficult. For example, at one site the waste quantity estimates list waste demilitarization protective ensemble suits separately, but at the other such waste is included in halogenated plastic waste.

**Recommendation 3-1.** The Program Executive Officer for Assembled Chemical Weapons Alternatives should consider implementing a uniform set of waste category designations for use at both the Blue Grass Chemical Agent Destruction Pilot Plant and the Pueblo Chemical Agent Destruction Pilot Plant to facilitate the transfer of knowledge and lessons learned between sites.

A typical DPE-suited operation is shown in Figure 3-2. The use of DPE suits is slow (limited to two workers at a time and no more than 2 hr per entry), cumbersome, physically tiring for the workers, and a major limitation for process throughput since operations are halted during the entries. Because of the human involvement, proximity to agent, and the awkward working conditions, DPE entries present challenges to safe and speedy work. A backup team of two workers is always standing by to assist with the safe egress of the workers. Plans are developed at each site that categorize the egress contingencies as follows:[6]

- *Normal cutout.* Most DPE entries are normal and consist of two workers in DPE suits doing a planned sequence of operations lasting no more than 2 hr. Before leaving the work area, they apply decontamination solution to their boots, gloves, knee area, or anywhere else they think may have come in contact with agent and walk to the airlock, allowing contact time for the decontamination solution to work. Once in the airlock, the workers rinse the decontamination solution off themselves and proceed to do "quadrant monitoring." One worker holds the sample point of a MINICAM near the other worker and moves it slowly over the other worker's body in a prescribed way quadrant by quadrant. The workers then switch roles. If less than 1 VSL is measured, then a normal cutout ensues and assistance is needed only to remove (cut off) the suits as they pass from one airlock to another.

- *Assisted cutout.* As workers are scanned in the airlock, if the readings are greater than 1 VSL, then some type of assistance is dictated. An example would be a worker who tried to decontaminate a boot or glove without success. In this case, an assistant would come into the airlock with appropriate personal protective equipment (PPE) gear, such as an M40 mask, gloves, and apron, to help with the cutout so that the worker(s) doffing the suit would not

---

[6]Personal communication between Harrison Pannella, study director, Jeff Kiley, Chief, Quality Assurance, CMA, and C.T. Anderson, Safety and Surety Engineer, ACWA, on October 26, 2011.

FIGURE 3-2 Workers in personal protective equipment working at a chemical weapons disposal facility. SOURCE: CMA Fact Sheet on Safe Disposal of Secondary Waste. Available online at www.cma.army.mil.

touch the outside of the suit where agent might still be present. The assistant comes in solely to help the workers doff their suits, not to help decontaminate. The aim is to help a worker avoid contact with the outer surface of the suit, where contamination may be present.

- *Hot cutout.* This type of egress occurs when there is a high level of residual contamination that cannot be successfully decontaminated. (For example, the agent might be embedded in grease on the suit.) An assistant wearing a higher level of PPE gear, such as a self-contained breathing apparatus along with gloves, boots, and other dermal protection, would enter the airlock to help remove the worker's suit.

- *Emergency cutout.* In an emergency cutout there is some medical urgency (e.g., heat stress) such that there is no time to decontaminate and verify. The cutting out in this fast-moving situation is done in an expedited way by a backup rescue team, which may consist of one worker in Level B PPE and another in Level C, or both workers in Level B. This team does not look for contamination; rather, they will do a gross decontamination with the emphasis on avoiding contact with the outside surface of the suit. When the call is made for the rescue team, the other DPE-suited worker will immediately hose down or start dropping buckets of decontamination solution on the DPE-casualty worker while in the work area to provide some benefit before the rescue team gets there to extract the stricken worker. Typically, one member of the rescue team will enter the area and work with the other DPE-suited worker to extract the DPE-casualty worker while the other member of the rescue team prepares to receive the stricken worker. For example, one litter or sled might go into the work area with one of the rescue team members while the other member prepares another litter or sled to which the casualty worker will be transferred for handoff to a medical team. In an emergency egress (medical, fire, etc.) the DPE-suited workers will not necessarily leave through an airlock but may use an alternative emergency exit door.

Clearly, these classifications become more problematic going down the list from the normal cutout to the emergency cutout. Each of the egress categories could be expedited by having a quick-response, real-time monitoring device capable of pinpointing areas of agent contamination and/or by verifying the lack of agent. Such a capability would help ensure the safety of workers. Scenario 3A in Box 3-1 describes how the new monitoring techniques might be of value.

---

**BOX 3-1**

**Scenario 3A: Improving Worker Safety During DPE Entries**

Worker safety is of paramount concern, and knowing as rapidly as possible where agent is or is not would facilitate smarter and safer DPE activities.

Faster quadrant scanning would speed normal DPE cutout time, and more accurate pinpointing of any residual agent on the DPE suit could guide workers to focus further decontamination on those spots. If further decontamination is not possible, the information could be used to avoid contaminated spots. Also, real-time monitoring of materials left in the wake of an extracted worker might help coworkers avoid contamination.

The use of portable ambient ionization instrumentation might allow faster assessment of agent presence for DPE activities and expedite safer worker egress.

---

DPE suits that have been exposed in Category A areas are assumed to be contaminated and require monitoring for agent contamination. That is typically accomplished by headspace analysis, as described in the ACWA Chemical Agent Monitoring Concept Plan (MCP) for personal protective clothing and equipment (PCE) (U.S. Army, 2011a, p. 59):

> To monitor decontaminated PCE, the PCE will be placed in a container or room and held for at least 4 hours at a minimum temperature of 21° C (70° F). The atmosphere (i.e., headspace) inside the container or room will be monitored for contamination via a technique applicable to the monitoring level of interest (VSL or 8-hour WPL) to verify that agent concentrations are below the applicable AEL before the PCE may be sent to the laundry facility. If agent concentrations are

> **BOX 3-2**
>
> **Scenario 3B: Enabling More Efficient DPE Entries**
>
> DPE entries into contaminated process areas place great physical strains on workers in DPE gear and also involve potential agent exposures. In addition, munitions processing must usually be suspended during entry operations. Opportunities to focus and expedite entry activities may lead to safer entries and shorter process interruptions and the generation of less DPE secondary waste.
>
> Portable ambient ionization instrumentation may allow quick and reliable determination of agent contamination locations and levels during DPE entries—for example, by determining if observed liquid deposits are agent or nonagent (water, oil), by clearly defining contaminated areas to focus decontamination actions and confirm their effectiveness, and by generally expediting effective decontamination activities that in turn would allow workers to downgrade the level of personal protection necessary to work in contaminated areas when appropriate.

detected above the applicable AELs, the PCE will be further decontaminated and remonitored.

PCE includes DPE suits. While headspace analysis is a proven and effective method for verifying whether DPE suits are clean, it is also time consuming.

Scenario 3B in Box 3-2 provides a scenario for possibly reducing the number of DPE suits used during process maintenance, agent changeover, and closure activities.

**Finding 3-2.** Any new monitoring method that could efficiently and reliably locate and quantify agent contamination may make decontamination activities more efficient by:

- Enabling faster identification of leaking munitions and decontamination of machinery, potentially reducing the number and/or duration of DPE-suited entries during normal plant operations, agent changeover periods, and closure activities;
- Reducing the total amount of secondary waste;
- Speeding waste disposal; and
- Minimizing worker exposure.

## CHANGEOVER OF AGENT DISPOSAL CAMPAIGNS AT BGCAPP

At BGCAPP, chemical agents will be destroyed in sequential campaigns beginning with GB, which due to its relative volatility presents the greatest risk, and then proceeding to VX. A decision is pending on whether all or only some of the mustard agent H projectiles should be destroyed by use of an explosive destruction technology

(EDT) instead of being processed through the main BGCAPP processes. During the changeover period from one agent to the next, the MINICAMS are reconfigured to monitor for the next agent to be destroyed. The facility and associated equipment and machinery must be completely decontaminated of the prior agent before operations can continue.

In preparation for changeover, occluded space teams (OSTs) are formed to identify occluded spaces that may harbor agent contamination. Common areas of occluded space include closed pipes, pump cavities, cracks in concrete, and caulking seals around equipment or concrete joints. Discussions and definitions of the types of occluded spaces that may occur in chemical weapons disposal facilities are presented in Box 3-3. Where possible, equipment having occluded spaces will be bagged for headspace analysis to verify contamination levels.[7]

If headspace analysis indicates contamination of a large piece of equipment, it might prove beneficial to have a more local, real-time probe (such as ambient ionization mass spectrometry for surface analysis) that could pinpoint the contaminated area for decontamination. Changeover operations typically take several months to complete and any reduction in the period of time necessary for changeovers would speed the overall disposal campaign. The committee believes it could be beneficial to identify situations in which bagging could safely be eliminated, or at least greatly reduced, by use of surface analysis by ambient ionization mass spectrometry. In this method, a wand would be waved over the surface of, for example, a contaminated machine or a cavity identified by an OST that would otherwise require bagging and headspace monitoring. Examples of how a real-time surface analysis instrument could speed changeover are given in Scenario 3C in Box 3-4 and Scenario 3D in Box 3-5.

**Finding 3-3.** A local, real-time agent monitoring system capable of monitoring surfaces might enhance the effectiveness of occluded space survey teams by identifying problematic occluded spaces and identifying other sources of contamination, possibly reducing the time necessary to conduct agent changeovers or facility closure.

## CLOSURE OPERATIONS

Once a chemical agent disposal facility has completed the disposal campaign operations for all agents stored at the site, it transitions into closure operations. This involves the decontamination and removal of equipment and decontamination and demolition of any contaminated infrastructure and buildings areas. The objective of closure is to decontaminate and safely demolish to ground level, in a manner that provides for the safety and protection of the workers, the public, and the environment, all of the buildings that were exposed to agent contamination (NRC, 2010). Such closure activities must satisfy regulatory permit requirements and other applicable federal, state, and local stipulations whether or not the area is destined to be returned to the public

---

[7]Gary Groenewold, member of the Committee to Review Secondary Waste Disposal and Regulatory Requirements for the Assembled Chemical Weapons Alternatives Program, "Sources and Amounts of Agent-Contaminated Wastes," presentation to the committee on February 23, 2011.

## BOX 3-3

### Definition and Classification of Occluded Spaces

The term "occluded space" is generally interpreted by the U.S. Army CMA as "a confined volume within a system, structure, or component that was exposed or potentially exposed to liquid agent and has the potential to contain any quantity of agent contaminated liquid."[1] Building on this definition, it is useful to classify occluded spaces into three broad types. Note that the following types do not reflect an official classification protocol by the U.S. Army but are presented for descriptive purposes in this report.

*Occluded space, Type I.* This category of occluded space is putatively the largest of the three categories. It represents areas that agent in liquid or vapor form could potentially penetrate and/or reside in that would hinder or preclude its detection by vapor screening methods (e.g., tenting). To avoid erroneous vapor screening measurements, systems, structures, and/or components are disassembled prior to VSL measurements to eliminate Type I occluded spaces before demonstrating that decontamination was successful. The derivative pieces following disassembly, for example, steel parts and laboratory glass, typically exhibit readily defined surfaces with smooth geometries.[2] Examples of Type I occluded space include seams, crevices, cracks, fasteners, threads, tubing, valves, and unsealed joints.

*Occluded space, Type II.* This category of occluded space represents process system or structural components containing materials with properties of porosity, miscibility, and/or chemical affinity for the liquid agent. In these cases, there is a reasonable probability that any exposure to high levels of gaseous agent, liquid agent, or agent contaminated liquids may have resulted in significant quantities of agent being adsorbed, absorbed, and/or trapped within the matrix of the material. This may result in misleadingly low values of agent vapor concentration when VSL measurements are performed. Typically, such materials are either incinerated (at CMA demilitarization facilities equipped with suitable furnaces) or stored prior to chemical neutralization on-site. Examples of Type II occluded spaces include wooden pallets, spent activated carbon, polymer gaskets, pump oil, lubricating oil, porous materials (including spill pillows), and agent miscible liquids.

*Occluded space, Type III.* The third category of occluded space is challenging to anticipate, foresee, and detect, as it represents potential occluded spaces in the process system or structural components that may have been present upon initial fabrication or construction or that became occluded at a later time, either in use during agent destruction campaigns, or even following decommissioning and shipment of the material off-site. These occluded spaces may arise from nonideal construction of the individual process system or structural components that remain undetected or are thought to be impervious, or they may not exist during plant systemization but arise later—for example, through material degradation by geological, environmental, or chemical forces. Although it is difficult to assess the extent of Type III occluded spaces (if any), it is nevertheless important to recognize their potential to occur and become exposed—for example, during decommissioning or deconstruction campaigns and, longer term, as sequestered wastes age, such as those decontaminated to level 3X or 5X (NRC, 2007), that are shipped off-site for long-term disposal (e.g., in landfills).

---

[1] J.M. Kilcy, 2011. "Closure Briefing," presented to the committee on February 23, 2011.
[2] NRC. 2007. *Review of Chemical Agent Secondary Waste Disposal and Regulatory Requirements*, Box 3-1 "U.S. Army Decontamination Metrics for Potentially Exposed Materials," p. 40.

> **BOX 3-4**
>
> **Scenario 3C: Process Area Occluded Space Surveys and/or Absorbed Agent Surveys During Changeover or Closure Activities**
>
> Unventilated agent vapor monitoring shows that a process area is contaminated. A portable ambient ionization instrument might be able to quickly interrogate and survey identified occluded spaces or suspected absorbed agent-contaminated materials, machinery, equipment, and plant structures, including concrete walls, floors, seams, and interfaces, to determine if agent reservoirs might be present and, if they are, guide and confirm focused decontamination efforts, thus reducing changeover or closure time.

> **BOX 3-5**
>
> **Scenario 3D: Complex Contaminated Demilitarization Machine Needs Decontamination at Agent Changeover or Closure Activities**
>
> Bagging and monitoring headspace levels is not practical for some equipment, materials, and machinery and may not be the most cost-effective approach for others. A portable ambient ionization instrument might be able to quickly survey occluded spaces and/or absorptive material components and identify the contaminated parts to direct focused decontamination or component removal to expedite treatment.

domain. These operations produce large amounts of secondary waste, as summarized in Tables 3-2 through 3-4. To minimize decontamination and to protect workers, a well-defined process is followed based on established AELs for the agents (see Table 2-1). This process includes the following:

- Maintaining and reviewing documented agent history. Based on lessons learned at prior closure operations, a careful record of all contamination events, spills, leaks, and the like is maintained during plant operation and used to identify contaminated areas by what is termed "generator knowledge."
- Selecting decontamination methods for each type of waste.
- Using OSTs to survey the facility and identify occluded spaces where agent may have accumulated. These include hidden spaces in piping, pumps, etc., that are not readily cleaned and decontaminated.
- Decontamination, monitoring, and dismantlement of equipment. Figure 3-3 shows tenting of a large piece of equipment in preparation for vapor monitoring.

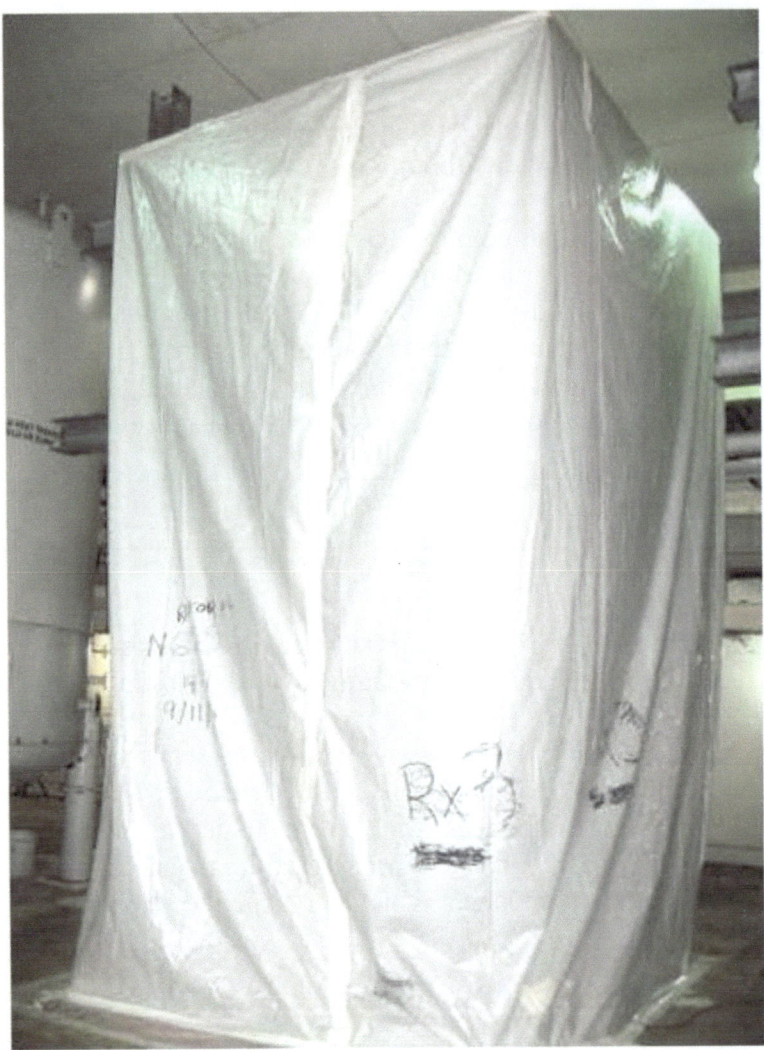

FIGURE 3-3 An example of a large item tented for monitoring at closure. SOURCE: Jeff Kiley, Chief, Quality Assurance, CMA, "CWD Closure Briefing," presentation to the committee on February 23, 2011.

- Decontamination and unventilated monitoring of the enclosed air space of building areas and rooms subject to having been contaminated by agent.
- Finally, razing buildings to the ground level.

Uncontaminated buildings may be retained based on agreement with the respective depot (Pueblo Chemical Depot or Blue Grass Army Depot) and any plans under base realignment and closure (BRAC) agreements. The criteria used to determine if an area is contaminated are contained in a U.S. Army review (2008a) and specified in the local environmental permits. Generally, the criteria areas are as follows:

- Any area or chamber exposed to liquid agent or aerosol is assumed to be contaminated.

- Any area or chamber exposed to agent vapor above the immediately dangerous to life or health (IDLH) level is considered to be contaminated.
- A decision on whether an area that has been exposed to agent vapor between IDLH and the short-term exposure limit (STEL) is contaminated is based on a risk assessment.
- Any area or chamber that has only been exposed to agent vapor below the STEL is assumed to not be contaminated.

Table 3-2 summarizes secondary wastes expected from closure operations at PCAPP and Tables 3-2 and 3-4 provide similar expectations for BGCAPP. Although the two sites use different terminology to describe some of their waste, much of the waste from normal operations, including from DPE suits and other toxicological agent protective (TAP) gear, continue to be major contributors during closure. In addition, closure creates significant streams of concrete, metals, combustible solids, and other incombustible solids. Also, all the activated carbon from the heating, ventilation, and air conditioning (HVAC) filter system must be disposed of at closure.

Special challenges to agent contamination monitoring are presented by some of these materials. For example, activated carbon is a strong absorber of agent, and concrete can be porous and can have cracks that accumulate agent. Combustible wood and fabric materials are often fibrous or porous, and agent may strongly adhere or absorb. Porosity can effectively provide large volumes of occluded space that are highly agent-retentive. As a lesson learned from prior chemical demilitarization sites, concrete that could potentially become exposed to liquid agent is of a high density and coated with a special polymer intended to minimize agent access to any cracks or pores. At BGCAPP and PCAPP, it is anticipated that this will minimize the need for scrabbling to remove agent-contaminated layers of concrete.

The current method of monitoring for agent adsorbed on or absorbed in concrete is headspace monitoring of nonventilated tented walls; this procedure includes heating to aid in vaporizing absorbed agent.[8] A previous NRC report questioned whether scrabbling would cause even more concrete to be classified as agent contaminated and recommended that ACWA investigate means for measuring residual agent on the concrete surfaces (NRC, 2008a, Recommendation 4-4). This committee concurs with that recommendation and suggests that ambient ionization mass spectrometry be investigated for that purpose. It now offers Scenario 3E (Box 3-6) as a possibility for further discussion in Chapters 4 and 5.

**Finding 3-4.** Materials with inherent porosity can readily adsorb or absorb agent and present a monitoring challenge for headspace vapor measurement methods.

---

[8]Personal communication from James Richmond, Director, ACWA Risk Management Directorate, to the committee, August 15, 2011.

> **BOX 3-6**
>
> **Scenario 3E: Concrete Waste Contamination Evaluation**
>
> Concrete that will have to be disposed of at closure may have agent adsorbed to its surfaces or absorbed into its mass. A portable ambient ionization instrument might be able to quickly and reliably interrogate adsorbed agent on the concrete surfaces and/or absorbed agent in bulk concrete to identify contaminated areas, estimate contamination levels, and reduce the need for scrabbling.

## ACTIVATED CARBON DISPOSAL

Activated carbon (also referred to as charcoal in Tables 3-1 and 3-2) strongly absorbs agent and other organics and is thus used in protective masks for workers and the final filter banks before air is released from the plant. The bulk of the activated carbon must be disposed of at closure. Over 100,000 lb are expected to be disposed of at BGCAPP (Table 3-5) and about 35,000 lb at PCAPP (Table 3-2).

Activated carbon is a porous adsorbent that cannot be verified to be free of agent by headspace analysis because the agent is strongly adsorbed on or absorbed in the carbon. Shipment of agent-exposed activated carbon to off-site disposal or recycling facilities requires verifying the mass of agent on the carbon. The method currently being pursued is solvent extraction of the adsorbed/absorbed phase from the carbon sample followed by GC/MS analysis. Applying this method at the Anniston Chemical Agent Disposal Facility (ANCDF) in Alabama, Southwest Research Institute found that VX was below the WCL but GB was above it. The GB values were later ascribed to the re-formation of GB from the hydrolysis products during extraction. A way was found to limit this re-formation, but if a real-time analysis technique were available that would not require extraction, it might prove to be a better alternative.

The NRC-recommended solution for the ultimate disposal of the activated carbon at ANCDF was to fill the polyethylene drums that held the carbon with caustic solution prior to transporting them to a treatment, storage, and disposal facility (TSDF) (NRC, 2009b). Scenario 3F in Box 3-7 asks whether surface ambient ionization mass spectrometry might be able to directly measure agent contamination on carbon, which might lead to more informed carbon disposal decisions.

> **BOX 3-7**
>
> **Scenario 3F: Spent Activated Carbon Contamination Evaluation**
>
> Spent activated carbon is likely to have bulk absorbed agent. A portable ambient ionization instrument might be able to quickly and reliably interrogate adsorbed/absorbed agent in spent activated carbon to identify contaminated materials, estimate contamination levels, and thus inform and focus decontamination or disposal activities.

## SCENARIOS SUMMARY

From a technical specifications perspective, the successful implementation of any new technologies to any of the scenarios presented in this chapter would represent a considerable departure from current agent monitoring practice, most noticeably in the detection of absorbed or chemisorbed materials on surfaces. In addition to directdetection on surfaces, a number of other analytical properties are critical to successful implementation for any particular application. These include purely analytical parameters such as sensitivity, dynamic range and selectivity as well as instrument factors such as reliability, portability, and ease of operation. Table 3-6 summarizes some of the criteria important to meeting the requirements of the different scenarios presented in this chapter. Chapters 4 and 5 will address the technical suitability of certain types of ambient ionization mass spectrometry for surface-adsorbed agent analysis.

TABLE 3-6 Critical Measurement Performance Criteria for Possible Scenarios

| Scenario | Detect on Surfaces | Dynamic Range | Portability | Precise Target Localization | Rapid Reconfiguration | Rapid Result | Scanning | Sensitivity | Short Duty Cycle | Specificity |
|---|---|---|---|---|---|---|---|---|---|---|
| 3A Improving worker safety during DPE entries | X | | X | | | X | | X | X | |
| 3B Enabling more efficient DPE entries | | | X | | X | X | | | X | X |
| 3C Process area occluded space surveys and/or absorbed agent surveys during changeover or closure activities | X | X | X | X | | X | X | X | | |
| 3D Complex contaminated demilitarization machine needs decontamination at agent changeover or closure activities | X | | | X | | | X | | | |
| 3E Concrete waste contamination evaluation | X | | X | | X | | | | | X |
| 3F Spent activated carbon contamination evaluation | X | | | | | | | X | | X |

# 4
# Current Status of Surface Measurement Technologies and Potential ACWA Site Applications

## INTRODUCTION

There are several overarching system requirements for the detection and monitoring of distilled mustard agent (HD) and distilled mustard mixed with bis(2-chloroethylthioethyl) ether (HT) at the Pueblo Chemical Agent Pilot Plant (PCAPP) and mustard agent (H) and nerve agents GB and VX at the Blue Grass Chemical Agent Pilot Plant (BGCAPP). These include an analytical system that (1) provides demonstrated reliability, (2) meets monitoring figures of merit[1] for the agents of interest at the desired concentrations in target matrices, (3) uses sampling intervals sufficient for protection of individuals and the environment, (4) conforms with accepted industrial hygiene principles and regulatory standards, and (5) allows sound statistical sampling selection to capture the exposure conditions or relevant waste contamination levels of the material matrices being characterized (U.S. Army, 2011a).

Distinguishing the degree to which various analytical systems can provide these capabilities for chemical agents on or within specific matrices depends on the ability of the analytical system to characterize different phases of the target species (i.e., gas, liquid, and solid) and the information content derived from such measurements. Levels of information content from measurements are described for a particular system through the monitoring method classification detailed in U.S. Army (2011a), whereby Class I methods provide quantitative and accurate values over the desired concentration range (using, for example, the depot area agent monitoring system [DAAMS]), Class II methods provide semi-quantitative and positive detection above a designated concentration limit (using, for example, the miniature continuous air monitoring system [MINICAMS]), and Class III methods, which are qualitative, provide only a binary positive or negative response for a specific agent at the designated response level (e.g., MINICAMS measurements of headspace VSL).

The current detection technologies selected for process and waste characterization are primarily based on vapor-phase detection, either in the context of (1) direct air sampling in the process stream or ambient environments, or (2) decontamination and decommissioning endeavors through vapor headspace analysis of materials contained within an enclosed space (U.S. Army, 2011a). Two primary technologies are integrated

---

[1]These include meeting required sensitivity, specificity, and response time.

into the design and construction of PCAPP and BGCAPP: DAAMS and MINICAMS. For example, the PCAPP facility has incorporated 137 DAAMS into the air monitoring system, typically with adjacent MINICAMS (Waybright, 2011).

Briefly, DAAMS preconcentrate air samples for preset sampling intervals using a sorbent filled-tube or sample loop, followed by agent desorption and gas chromatography-mass spectrometry (GC-MS) characterization. Typically, samples are preconcentrated for an 8-hr collection time and a full analysis is performed within 24 hr to provide archival analytical results. In contrast, MINICAMS utilize cycled GC measurements to provide near-real-time (NRT) analyses that take approximately 5 to 15 min. Depending on the specific contaminated matrices expected during agent processing, changeover, or closure activities, the specific protocols used primarily conform with those prescribed by the U.S. Environmental Protection Agency.

The samples derived from the anticipated waste streams, including from plant decommissioning activities, vary widely in their physical characteristics (for instance, their state and volume and type of matrix) and the agent contamination spatial resolution desired. Although DAAMS and MINICAMS technologies for vapor analyses have been demonstrated to be reliable, they do not provide good spatial or temporal resolution and are not particularly well suited for the direct analysis of agents in condensed-phase wastes and in occluded or potentially occluded volumes (see Box 3-3).

Recent advances in analytical technologies, as described below, are anticipated to complement existing agent monitoring strategies and may improve workplace and environmental safety. They provide the ability to detect agent contamination directly for a variety of matrices as well as agent in the gas phase, potentially providing high spatial resolution and rapid temporal assessments of agent contamination. These features are particularly relevant for more efficient and timely characterization of wastes during the agent processing and agent changeover phases of demilitarization operations. The ability to rapidly scan protection gear (demilitarization protective ensemble (DPE) suits) as workers exit Class A areas, where the suits may have been contaminated by exposure to chemical agents, could enhance worker safety while greatly reducing the time spent in decontamination and transition. In addition, the committee envisions that the application of these advanced technologies during plant decommissioning (closure) activities could reduce the time required for this phase of the project, yielding significant cost savings. Finally, the spatial resolution and rapid temporal response of the new analytical methods described below might prove to be an invaluable asset in dealing with unanticipated events such as an agent release, facilitating the tracking of any vapor releases to their source.

Two recently developed surface measurement technologies based on mass spectrometry ambient ionization techniques have been commercialized and found widespread application: direct analysis in real time (DART) (Cody et al., 2005) and desorption electrospray ionization (DESI) (Takáts et al., 2004). These are illustrated in Figures 4-1 and 4-2 and discussed in greater detail below. DART, DESI, and related techniques (Harris et al., 2011) allow direct sampling and analysis of a wide variety of liquid or solid matrices and produce high spatial resolution information to localize the target analytes. DAAMS and MINICAMS provide information about time-averaged agent concentration levels in the air or from the vapor headspace of samples in enclosed spaces. Such information represents indirect measurements of where the contamination is

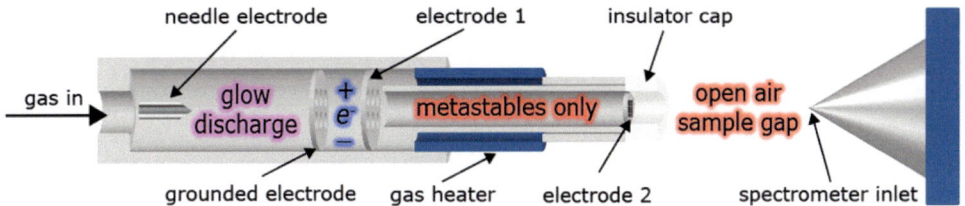

FIGURE 4-1 Schematic diagram of DART ion source. An electrical discharge creates a glow discharge plasma as inert gas flows through the DART chamber. The plasma contains ions, electrons, and excited neutral atoms and molecules. Biased electrodes remove ions and electrons, leaving only long-lived electronically or vibronically excited atoms and molecules (metastables). The gas can be heated to assist desorption of neutrals from surfaces. The heated gas exits the source heading toward the atmospheric pressure inlet of the mass spectrometer. The ions produced in the vapor phase between the DART insulator cap and the mass spectrometer inlet are formed by interaction of the electronic excited-state species of helium or neon or vibronically excited nitrogen with the sample. Samples are placed along the path between the cap and inlet. For larger samples the source is mounted at an angle with respect to the MS inlet.

SOURCE: JEOL USA, Inc., R.B. Cody. Reprinted with permission.

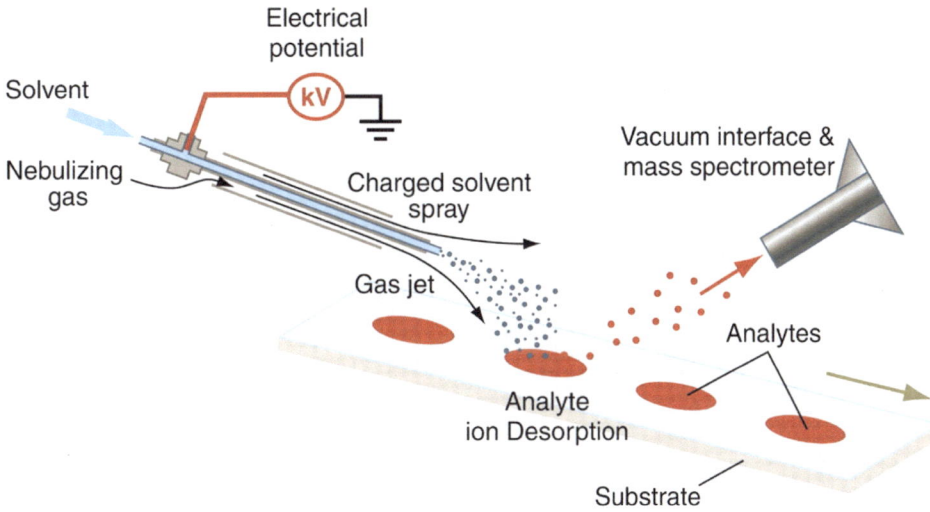

FIGURE 4-2 Schematic diagram of DESI ion source. A nebulizing gas flows around an electrospray source, which is directed at the surface to be analyzed. Charged droplets pick up analyte molecules from the surface and are captured by the vacuum interface of the mass spectrometer through a capillary of variable length. SOURCE: Cooks et al., 2006. Reprinted with permission from AAAS.

localized. In contrast, real-time, spatially resolved techniques such as DART and DESI provide the ability to allow directly identifying the location of agent contamination on solid matrices and, with appropriate modifications, DART may also be used to measure trace components in vapor plumes and locate their sources by tracking airborne agent concentration gradients in real time. Detailed descriptions of these recently developed ambient mass spectrometry technologies are provided later in this chapter.

The ion source is an integral part of most laboratory mass spectrometric instrumentation and is located in close proximity to the mass analyzer in order to maximize ion transport from the source to the analyzer. As a result, the operator must take the sample to the instrument. However, with the new sampling technologies it is possible to perform nonproximate analysis, in which the ion source is some distance (0.5 to 3 m) from the atmospheric pressure inlet, with ions transported through flexible tubing and sampled by remote wands. This approach enables the instrument to be brought to the sample and facilitates the use of a cart-mounted or even personnel-carried instruments capable of identifying localized regions of agent contamination on surfaces. In addition, even though DART and DESI are designed to detect surface species, it might be possible to further develop methods of operating DART and DESI instruments, or combine them with a second ambient ion source specifically for detection of trace species in the vapor phase, producing a single portable instrument able to perform real-time vapor and surface analyses.

In summary, the combination of real-time measurements with multi-state (gases, liquids, and solids) analytical capability suggests that the newer instruments for ambient ionization mass spectrometry incorporating remote sampling capability deserve serious consideration for their potential to supplement the current DAAMS and MINICAMS agent monitoring instruments. Based on the capabilities reviewed in this chapter and potential measurement strategies reviewed in Chapter 5, their potential to enhance workplace safety, improve operational efficiency, and accelerate decontamination activities during both operational and closure activities at the BGCAPP and PCAPP facilities will be assessed.

## PROPERTIES OF THE TARGET MOLECULES RELEVANT TO THEIR DETECTION BY AMBIENT MASS SPECTROMETRY

Before examining details of recent advances in ambient mass spectrometry, it is useful to review the ionization schemes and mass spectrometer configurations that are generally used to detect chemical agents. In this section, the properties of the target chemical species GB, VX, and mustard agents are reviewed in relation to their detection by ambient ionization mass spectrometry. This analysis also provides insights into possible interferences from other trace species in the sampled environment that could compromise the detection of the target chemical agents.

The best way to detect a molecule by mass spectrometry is to leave the molecule intact through a soft ionization process—that is, a process that is not energetic enough to fragment the resulting ion. Such soft processes are often chemical ionizations that are only mildly exothermic, leaving products with insufficient internal energy to fragment. These processes are in contrast to electron impact ionization, in which high-energy electrons excite and ionize the molecule, which then fragments readily. This can lead to ambiguity in the identification of the precursor, especially if several neutral species are simultaneously ionized in a mixture. In addition, chemical ionization, unlike electron impact ionization, seldom produces highly reactive and easily fragmented odd electron species.

In the ambient ionization instrumentation discussed later in this chapter, a variety of classes of chemical ionization reactions are likely depending on how the source is configured and operated. Electron transfer involves a precursor ion accepting or donating an electron to the molecule to be ionized. The reaction of a chemical agent, C, with a reagent ion, $A^+$ or $A^-$, may be illustrated as follows:

For a positive chemical agent ion, $A^+$:

$$A^+ + C \rightarrow C^+ + A \qquad (4.1a)$$

For a negative reagent ion, $A^-$:

$$A^- + C \rightarrow C^- + A \qquad (4.1b)$$

Charge transfer produces an ion with the same mass as the chemical agent, which is obviously desirable. The important parameters in determining whether reaction 4.1a or 4.1b can occur are ionization potentials (IP) for positive ions and electron affinities for negative ions. The former ionization potential is defined (Lias et al., 1988) as the energy at 0 K for the process in reaction 4.2:

$$C \rightarrow C^+ + e^- \qquad (4.2)$$

Reaction 4.1a may then occur if the ionization potential of the chemical agent is less than the ionization potential of the parent molecule of the ion precursor, A. In dry air, a typical reagent ion may be $NO^+$. The important property for negative ion charge transfer is the electron affinity, defined by Lias et al. (1988) as the negative of the 0 K energy for attaching an electron, as illustrated in reaction 4.3:

$$C + e^- \rightarrow C^- \qquad (4.3)$$

In this case, reaction 4.1b occurs if the electron affinity of the chemical agent is greater than that of the reagent ion $A^-$.

Proton donors are the most commonly used ionization reagents in ambient ionization mass spectrometry. In air with even a modest amount of humidity, a well-known series of reactions take place that rapidly form protonated water, $H_3O^+$, and its hydrates (Ferguson and Fehsenfeld, 1969). Just as its name implies, proton transfer involves moving a proton from the ionizing agent to the chemical agent, as illustrated by reaction 4.4:

$$H_3O^+ + C \rightarrow H^+C + H_2O \qquad (4.4)$$

Here the important property in determining the likelihood of reaction is the proton affinity, defined as the negative of the energy at 298 K of adding a proton to a molecule (Lias et al., 1988), as expressed by reaction 4.5:

$$H^+ + A \rightarrow HA^+ \tag{4.5}$$

Reaction 4.4 will most likely occur if the proton affinity of the chemical agent is greater than that of the $H_3O^+$ reagent ion (Bohme, 1975). In the vast majority of exothermic proton transfer reactions, the rate constant is very close to the collisional limit (Anicich, 2003; Su and Chesnavich, 1982). Ambient ion mass spectrometry (like all chemical ionization methodologies) derives its sensitivity from the fact that the ion–molecule collision rate is large. After proton transfer, the mass of the chemical agent in daltons is one number greater than its molecular weight. In sources involving clean air with at least a little water, proton hydrates are the precursor ions unless trace impurities with high proton affinities can sequester the labile proton before it reacts with the target chemical agent (Ketkar et al., 1991a). For example, $NH_3$, present in exhaled breath, can act in this way. In this case, $NH_4^+$ and its hydrates are the precursors. These will be more selective ionization agents since $NH_3$ has a considerably higher proton affinity than does $H_2O$ and therefore transfers a proton to many fewer molecules than does $H_3O^+$ (Linstrom and Mallard, 2007). To avoid situations where atmospheric variation in $NH_3$ affects detection conditions, a trace amount of $NH_3$ may be added to the source to provide consistent performance. Other dopants may be added to fine tune the energetics of proton transfer and further optimize the detection selectivity. There is an equivalent negative ion reaction that is useful for the detection of strong acids. This involves reaction of a proton acceptor such as $OH^-$ or $O_2^+$ with the target molecule to generate the conjugate base by proton transfer. Since the chemical agents of interest are not strong acids this is not discussed further.

The above reactions detect the agents at either their neutral mass or at 1 dalton higher. There are alternative chemical ionization schemes that involve larger changes in mass. A common one in the atmospheric community involves $F^-$ transfer from an appropriate fluoride donor reagent ion. In a prior NRC report on chemical agent monitoring, it was suggested that this class of reaction might be extended to detect chemical agents (NRC, 2005a). Since that time, it has been shown that the fluoride ion affinities of HD, GB, and VX are low, so this process will not be useful (Midey et al., 2008, 2009, 2010).

Studies involving simulants and recent work on agents themselves have shown that clustering is likely to be important (Nilles et al., 2009). Here the chemical agent and reagent ion will simply stick together, as indicated in reaction 4.6,

$$A^+ + C + M \rightarrow (AC)^+ + M \tag{4.6}$$

where M is an atmospheric molecule that acts as a third body to remove excess energy from the adduct, leading to formation of a stable complex. The work of Nilles et al. (2009) using DART showed that $NH_4^+$ clustered to the G series of agents.

All of the above assumes that no trace species is in large enough concentration to deplete the primary ion significantly. In that case, the kinetics becomes more complicated. The product ions are no longer proportional to concentration. Additionally, one may need to account for reactions of the product ions. Since the interest is to detect contaminants and not large spills, neglecting these effects should not affect the ability to detect an agent.

In detecting chemical agents it is clear from the above discussion that it is extremely important to understand their properties. Table 4-1 extends Table 2-1 from a prior NRC report (NRC, 2005a). It lists a variety of chemical properties for GB (sarin), VX, HD, and HT, including vapor pressure and several ion energetic properties. Except for GB, all the agents of interest have exceedingly low vapor pressures.

The ion energetics properties shown in Table 4-1 indicate that a number of positive chemical ionization reagents should readily ionize the target chemical agents. Electron transfer from an agent molecule to the reagent ion will occur for the common reagent $NO^+$ for VX and H agents since their IPs are lower than that of NO (IP = 9.26 eV) (Linstrom and Mallard, 2007). Since the IPs of the agents are close to that of NO, little fragmentation is expected. Experiments on surrogates confirm this (Midey et al., 2008, 2010). On the other hand, GB has an IP higher than NO, and experiments on surrogates indicate that clustering is likely (Midey et al., 2009). Higher-energy charge transfer ions, such as $O_2^+$, will react readily with agents but will lead to fragmentation of the agents, making identification more difficult (Midey et al., 2008, 2009, 2010).

As discussed above, most chemical ionization schemes center on proton transfer. $H_3O^+$ is readily formed in ambient mass spectrometry and is the reagent ion most likely to transfer a proton since $H_2O$ has a low proton affinity (691 kJ/mol) (Linstrom and Mallard, 2007). All agents have higher proton affinities than $H_2O$ and therefore proton transfer is highly probable. This is confirmed in laboratory studies on agent surrogates that show that proton transfer occurs. However, due to the amount of energy available, dissociative proton transfer may also occur, and work on agent surrogates indicates that this may happen (Midey et al., 2008, 2009, 2010). Therefore, it is better to use reagent ions that provide less exothermic proton transfers. $NH_3$ has a proton affinity of 854 kJ/mol (Linstrom and Mallard, 2007). This is considerably lower than the proton affinity of VX, equal to that of GB within error margins, and larger than that of mustard agent. Indeed, the surrogate experiments show that proton transfers from $NH_4^+$ to VX and GB are likely and that mustard agent clusters to $NH_4^+$ (Midey et al., 2008, 2009, 2010). The proton affinity of VX is near the top of the scale, and almost any protonated reagent ion will ionize it. A reagent ion with a high proton affinity, but slightly less than that of VX, is likely to be very selective by reducing background from interferences and would therefore permit detection of VX with high sensitivity.

Negative ions show much less promise for detection of the chemical agents relevant to ACWA operations. All of the agents have exceedingly low electron affinities, and therefore few parent ions will be formed. In addition, all of the agents have fairly low fluoride affinities. A commonly used reagent ion is $SF_6^-$. The fluoride affinity of $SF_5$ is 230 kJ/mol, which is much higher than that of any of the chemical agents. It is possible that some other fluoride transfer reagent might be more appropriate, but there is no obvious choice at present.

The above discussion emphasizes that soft ionization techniques are likely to ionize a variety of agents without fragmentation. However, that is not sufficient for conclusive identification of chemical agents. The chemical environment in

TABLE 4-1 Physical Properties of Chemical Warfare Agents

| Agent Characteristic | Nerve Agents | | Blister (Mustard) Agents[a] | |
|---|---|---|---|---|
| | GB (Sarin) | VX | HD | HT |
| Chemical formula | $(CH_3)_2CHO(CH_3)FPO$ | $C_{11}H_{26}NO_2PS$ | $(ClCH_2CH_2)_2S$ | 60% HD, 40% $((ClCH_2CH_2)_2SCH_2CH_2)_2O$ |
| CAS Registry No. | 107-44-8 | 50782-69-9 | 505-60-2 | N/A |
| Molecular weight | 140.10 | 267.38 | 159.08 | (Mixture – 188.96 based on 60/40 weight percent) |
| Boiling point (°C) | 150 (extrapolated) | 292 (extrapolated) | 218 | No constant boiling point |
| Freezing point (°C) | –56 | <–51 | 14.5 | 0 to 1.3 |
| Vapor pressure (mm Hg at 25 °C) | 2.48 | 0.0009 | 0.106 | $7.7 \times 10^{-2}$ (calculated based on Raoult's law equation and 60 weight percent HD and 40 weight percent T) |
| Vapor density (relative to air) | N/A | N/A | 5.5 (calculated) | 6.5 (calculated based on 60/40 weight percent) |
| Volatility (mg/m$^3$) | 3,370 at 0 °C  187,000 at 25 °C | 12.6 at 25 °C | 75 at 0 °C (solid)  906 at 20 °C (liquid) | 783 at 25 °C |
| Surface tension (dyne/cm) | 26.5 at 20 °C | 32.0 at 20 °C | 43.2 at 20 °C  42.4 at 25 °C[b] | 42.0 at 25 °C |
| Kinematic viscosity (cS) | 1.28 at 25 °C | 12.26 at 20 °C | 3.52 at 20 °C | 6.50 at 20 °C |
| Liquid density (g/cm$^3$ at 20 °C) | 1.0887 | 1.0083 | 1.27 | 1.26 |
| Latent heat of vaporization (cal/g) | 82.9 | 71.8 | 94.3 | N/A |
| Solubility (g/L H$_2$O at 25 °C) | Completely miscible | 50 at 21.5 °C | 0.92 | Similar to HD |
| Heat of combustion, (cal/g) | 5,600 | 8,300 | 4,500 | N/A |
| Ionization potential (eV) | 9.82 | 7.3 | 8.74 | N/A |
| Proton affinity (kJ/mol) | 857 | 1,039 | 796 | N/A |
| Fluoride affinity (kJ/mol) | 152 | 111 | 104 | N/A |

[a] The blister agents are labeled H, HD, and HT. The active ingredient in all these agents is bis(2-chloroethyl) sulfide, $(ClCH_2CH_2)_2S$. HD, called distilled mustard, is nominally pure bis(2-chloroethyl) sulfide. H, often called Levinstein mustard, is approximately 70 percent bis(2-chloroethyl) sulfide and 30 percent impurities, which tend to be polysulfides such as $(ClCH_2CH_2)_2S_n$. HT is a mixture of ca. 60 percent $(ClCH_2CH_2)_2S$ and 40 percent $((ClCH_2CH_2)_2SCH_2CH_2)_2O$.

[b] The surface tensions of HD at both 20°C and 25°C are included to allow the reader to compare the surface tensions of HD and HT under the same physical conditions, while also giving the reader a general comparison of the surface tensions of nerve agents and blister agents across constant physical conditions.

SOURCE: NRC, 2005b, 2005c; Midey et al., 2008, 2009, 2010.

demilitarization plants is likely to be complex, and a wide variety of trace species are likely to be present. If one or more of these species has the same nominal mass as the agent when ionized, a false positive could occur with low-resolution mass spectrometric detection. In order to separate the ions derived from agents from those due to other trace species, two techniques are useful. While it may be common to have two species (agent and other) at the same integer (nominal) mass (e.g., CO and $N_2$), it is unlikely that they will have the same exact mass. Suitably compact mass spectrometers with high enough mass resolution to determine exact masses are now common, although more expensive than lower resolution counterparts. A mass spectrometer with a 5 part per million mass resolution is able to identify elemental composition for species up to 200 daltons.

An alternative strategy involving less expensive instrumentation is to isolate ions with the same nominal mass and then fragment them using tandem mass spectrometry (MS/MS). After the ions with the same nominal mass are isolated, they are accelerated and collide with a nonreactive species (e.g., He or Xe), causing fragmentation. Different species with the same nominal mass will have different fragmentation patterns. Fragmentation often occurs to form not one but many ions, as indicated in reaction 4.7:

$$A^+ + M \rightarrow B^+, C^+, D^+ \ldots + M \qquad (4.7)$$

where M is the unreactive collision partner used to transfer the energy needed for fragmentation. The mass spectrum of the fragments ($B^+$, $C^+$, $D^+$), which is called the product ion spectrum, determines which species is present. Tandem mass spectrometry can be arranged to occur either in time or in position. Quadrupole traps are common for the former and sequential instrumentation, such as triple quadrupoles, are used for the latter. As an example, a laboratory at the U.S. Army's Edgewood Chemical Biological Center has used such a scheme to determine VX levels in animal organs. The extract is ionized in a GC-MS equipped with an $NH_4^+$ chemical ionization source. Protonated VX is isolated at 268 daltons. However, there are other compounds in the extract with this mass. The 268-dalton ion is passed through a collision cell and the fragment ion at 128 daltons is identified as coming only from VX (McGuire et al., 2008).

Published chemical ionization measurements of chemical agents confirm the ion reaction mechanisms discussed above. Atmospheric pressure ionization MS/MS measurements using protonated water vapor reagent ions by Fite and coworkers (Ketkar et al., 1991b) yielded detection limits for GB of 14.1 parts per trillion by volume ($ppt_v$), and detection of VX fragments cleaved to produce the G analog by reaction with a silver fluoride impregnated pad yielded a detection limit of 100 $ppt_v$. More recently, Monks and coworkers (Cordell et al., 2007) demonstrated real-time detection of GB and H at detection levels of 3-5 parts per billion by volume ($ppb_v$), also using protonated water vapor as the reagent ion.

# EXPERIMENTAL METHODS FOR AMBIENT MASS SPECTROMETRY

Ambient, or open air, surface sampling techniques, sometimes described collectively as "ambient MS techniques," are a group of ion generation approaches that can be readily coupled to mass spectrometric detectors for both MS and tandem MS (MS/MS) analysis of target molecules. These technologies have been coupled to a variety of mass spectrometers equipped with atmospheric pressure interfaces with only minor modifications, enabling the identification of unknowns by fragmentation pattern matching to databases, elemental formula determination via accurate mass measurements, multianalyte quantitation, spatially resolved measurements, and selective ionization enhancement for target compounds of interest. Ambient MS sampling/ionization techniques such as DART and DESI have grown in popularity because they:

- Enable the sampling of analyte under atmospheric pressure conditions from both liquid and solid states remotely from the mass analyzer.
- Have been used to investigate objects or surface features of a wide range of shapes, sizes and textures.
- Can perform qualitative or quantitative analysis with no or minimal sample preparation, such as dissolution, grinding, extraction, or preconcentration.
- Can conduct all these operations in real time with high sensitivity and minimal unwanted ion fragmentation.

Although this chapter focuses mainly on the two most mature ambient sampling approaches (DESI and DART), a multitude of other approaches have been described in the last 5 yr. Table 4-2 (Harris et al., 2011) lists acronyms for different ambient MS techniques and associated references. This list shows the explosive growth of this field. Schematic illustrations of the experimental methodologies associated with these ambient MS techniques are shown in Figure 4-3 through Figure 4-6. The great variety of techniques reported can be classified into the following broad categories:

- Sampling techniques based on spray and solid-liquid extraction, where ionization proceeds by electrospray ionization (ESI)-like mechanisms,
- Direct and alternating current plasma techniques, where ionization proceeds via chemical ionization mechanisms and sampling proceeds via chemical sputtering or thermal desorption,
- Techniques involving laser desorption or ablation followed by ESI or plasma postionization, and
- Acoustic volatilization methods that involve electrospray or chemical ionization after volatilization.

In terms of technology maturity, only DESI and DART have been commercially available for several years. More recently, laser ablation electrospray ionization (LAESI) and liquid extraction surface analysis (LESA) have also become commercially available. DART and DESI have been by far the most popular techniques in terms of the range of demonstrated applications, the number of users, and the number of scientific publications.

TABLE 4-2 List of Acronyms (Ordered Alphabetically) and Relevant References Describing Various Ambient Surface Sampling Techniques

| Acronym | Name | First Report |
|---|---|---|
| **Spray and solid-liquid extraction based** | | |
| DAPPI | Desorption Atmospheric Pressure Photo Ionization | Haapala et al., 2007 |
| DESI | Desorption Electrospray Ionization | Takáts et al., 2004 |
| DICE | Desorption Ionization by Charge Exchange | Chan et al., 2010 |
| EASI | Easy Ambient Sonicspray Ionization | Haddad et al., 2006 |
| LESA | Liquid Extraction Surface Analysis | Kertesz and Van Berkel, 2010 |
| LMJ-SSP | Liquid Micro Junction-Surface Sampling Probe | Wachs and Henion, 2001 |
| ND-EESI | Neutral Desorption Extractive Electrospray Ionization | Chen et al., 2007 |
| PESI | Probe Electrospray Ionization | Hiraoka et al., 2007 |
| **Plasma-based** | | |
| DAPCI | Desorption Atmospheric Pressure Chemical Ionization | Takáts et al., 2005a |
| DART | Direct Analysis in Real Time | Cody et al., 2005 |
| DBDI | Dielectric Barrier Discharge Ionization | Na et al., 2007 |
| DCBI | Desorption Corona Beam Ionization | Wang et al., 2010 |
| FAPA | Flowing Atmospheric Pressure Afterglow | Andrade et al., 2008 |
| LTP | Low-Temperature Plasma probe | Harper et al., 2008 |
| **Laser desorption/ablation based** | | |
| ELDI | Electrospray-assisted Laser Desorption Ionization | Shiea et al., 2005 |
| IR-LAMICI | Infrared Laser Ablation Metastable-induced Chemical Ionization | Galhena et al., 2010 |
| LADESI | Laser-Assisted Desorption Electrospray Ionization | Rezenom et al., 2008 |
| LAESI | Laser Ablation Electrospray Ionization Mass Spectrometry | Nemes and Vertes, 2007 |
| LDESI | Laser Desorption Electrospray Ionization | Sampson and Muddiman, 2009 |
| MALDESI | Matrix-Assisted Laser Desorption Electrospray Ionization | Sampson et al., 2006 |
| **Acoustic-based** | | |
| LIAD-ESI | Laser-Induced Acoustic Desorption-Electrospray Ionization | Cheng et al., 2009 |
| RADIO | Radio-frequency Acoustic Desorption and Ionization | Dixon et al., 2009 |
| **Other** | | |
| AP-TD/SI | Atmospheric Pressure Thermal Desorption/Secondary Ionization | Basile et al., 2010 |
| BADCI | Beta electron-assisted Direct Chemical Ionization | Steeb et al., 2009 |
| DEMI | Desorption Electrospray/Metastable-Induced Ionization | Nyadong et al., 2009 |
| REIMS | Rapid Evaporative Ionization Mass Spectrometry | Schäfer et al., 2009 |
| SwiFerr | Switched Ferroelectric Plasma Ionizer | Neidholdt and Beauchamp, 2011 |

SOURCE: Adapted from Harris et al., 2011.

Coupling of these front-end techniques to detectors providing tandem MS and/or accurate mass capabilities has been shown to significantly improve sensitivity and specificity, as defined in Chapter 5. Detailed descriptions of DART, DESI, and a few closely related approaches are presented in the following sections. While the focus of this review is on surface sampling, it is noteworthy that many of the ambient sampling techniques can directly ionize trace species and provide means for their detection in the vapor phase with high sensitivity.

## Direct Analysis in Real Time (DART)

Plasma-based ambient sampling techniques such as DART, FAPA, LTP, DBDI, DCBI and DAPCI (see Table 4-2 for definitions) involve the generation of a direct-current or radio-frequency electrical discharge between a pair of electrodes in contact with a flowing support gas such as $N_2$ or He, generating a constrained flux of ions, radicals, excited-state neutrals, and electrons, which ultimately lead to ionization of the sample. Some or all of these plasma species can be directed toward the surface being sampled, inducing desorption and ionization in a single step. Optional resistive heating of the support gas can further enhance desorption of neutrals, which are then ionized by interaction with plasma species. Plasma-based ambient MS instrumentation tends to be fairly simple and rugged and can be coupled to a variety of mass spectrometers, including, most commonly, quadrupole ion traps, linear quadrupole ion traps, and quadrupole time-of-flight analyzers, providing MS/MS and/or accurate mass-determining capabilities for DART-produced ions. Plasma source mass spectra tend to be relatively simple, because most of the time the analytes are ionized, as described in the previous section, as one or two adduct types, simplifying peak assignment in the case of unknowns. Their applicability is generally limited to molecular weights below 1 kDa. Chemical agent masses are well below this limit.

DART (Figure 4-1) uses a point-to-plane atmospheric pressure glow discharge to generate metastable species in a chamber that is physically separated from the ionization region. The discharge support gas, containing metastables, is heated and directed through a grid electrode that filters ions and electrons to mitigate ion-ion and ion-electron recombination of species generated within the DART ionization source. DART can be used to sample gases, liquids, and solids. In laboratory settings, gases are directly injected into the ionization region following the grid electrode, whereas liquids are generally sampled by dipping a glass capillary in the sample and placing it in the ionization region. Solids can be directly analyzed by being exposed to the stream of ionizing gas, or conveyed in a transmission mode geometry, where, for example, the sample may be coated on an open mesh (Perez et al., 2010). Powders can be mixed with metal particles, adhered to a permanent magnet, and exposed to the DART gas. Foam swabs (Edison et al., 2011), solid-phase extraction materials (e.g., adsorbent fibers) (D'Agostino and Chenier, 2010), and polydimethylsiloxane-coated stir bars (Haunschmidt et al., 2010) can also be directly placed within the DART ionization region. Desorption can also be achieved via infrared (IR) laser ablation of the surface being examined, or IR-LAMICI, resulting in higher spatial resolution (Figure 4-5) (Galhena et al., 2010).

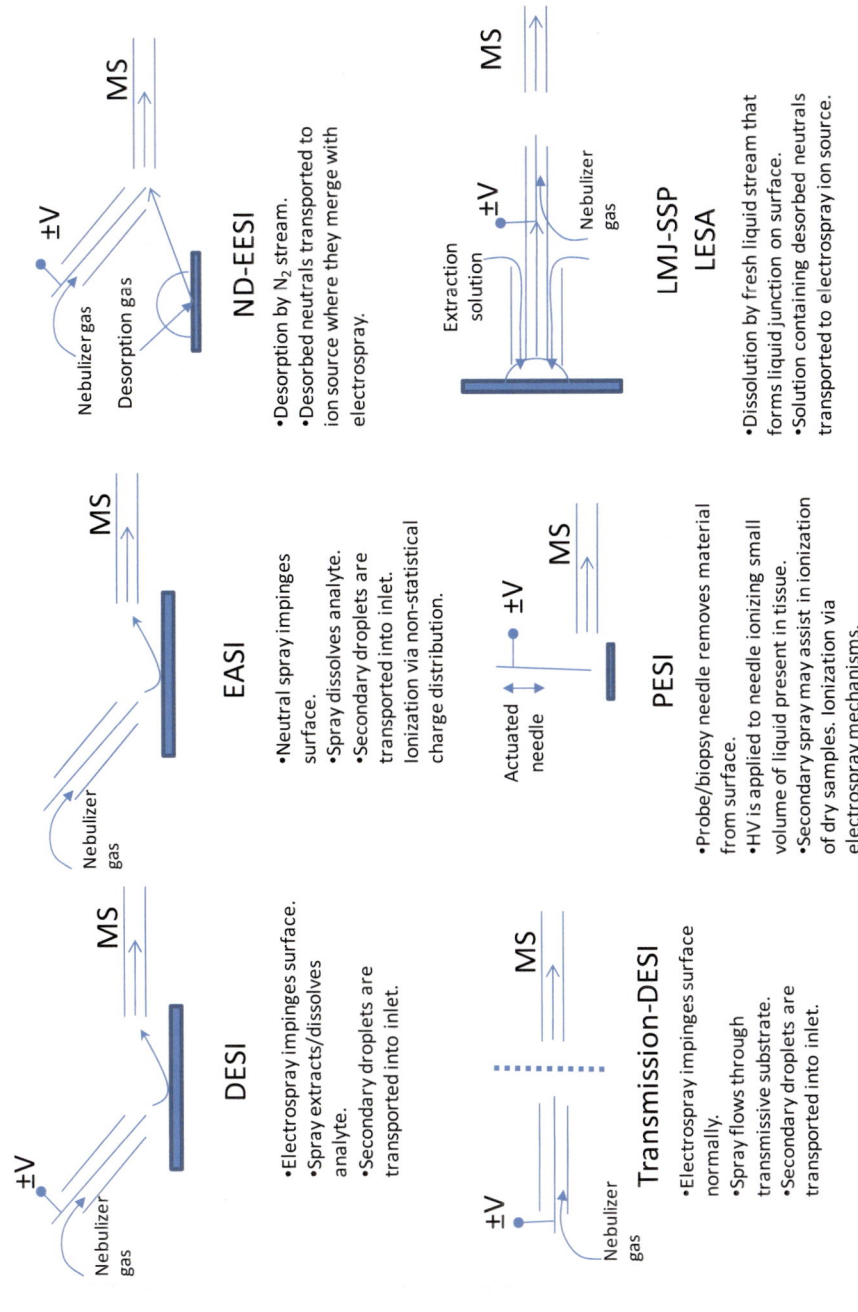

FIGURE 4-3 Schematic illustrations showing the operation of several different ion sources and sampling schemes for ambient mass spectrometry. NOTE: HV = high voltage. SOURCE: Harris et al., 2011. Reprinted with permission. Copyright 2011 American Chemical Society.

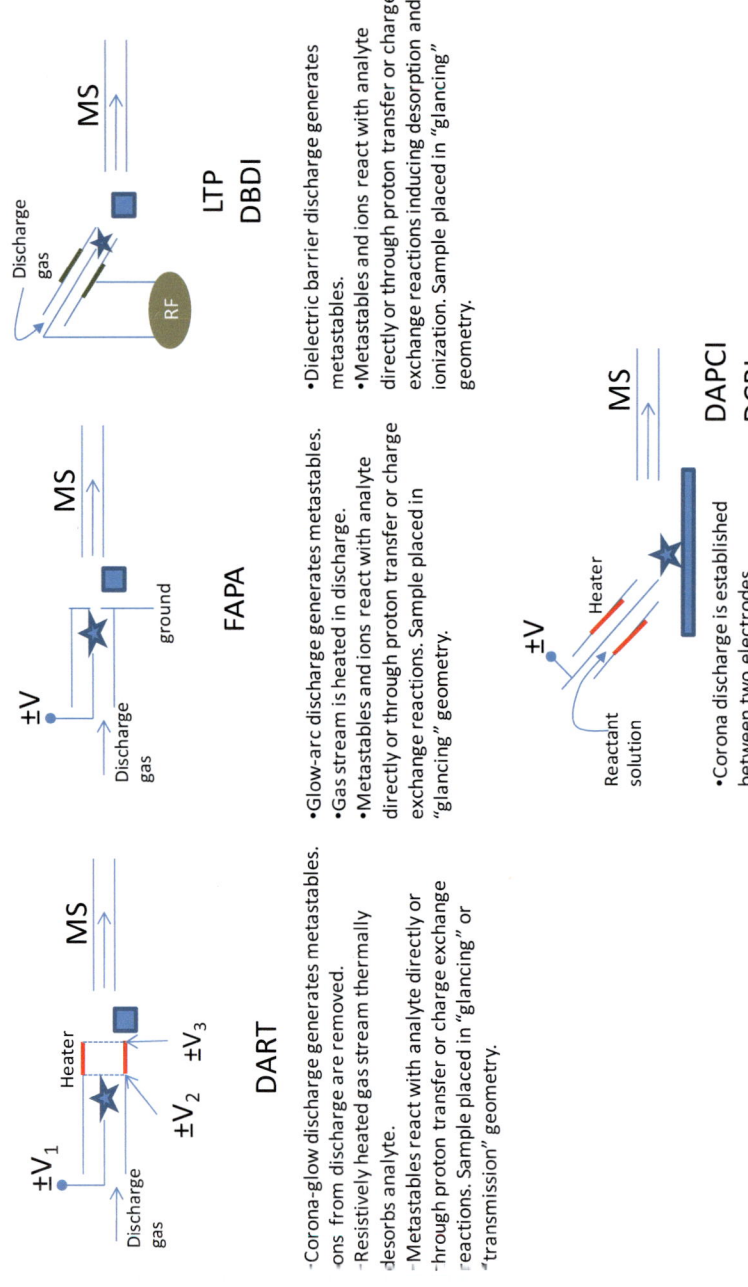

FIGURE 4-4 Additional illustrations showing the operation of several different ion sources and sampling schemes for ambient mass spectrometry. SOURCE: Harris et al., 2011. Reprinted with permission. Copyright 2011 American Chemical Society.

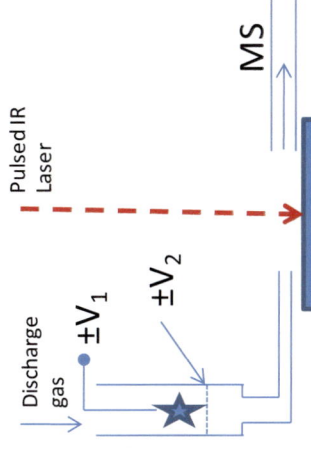
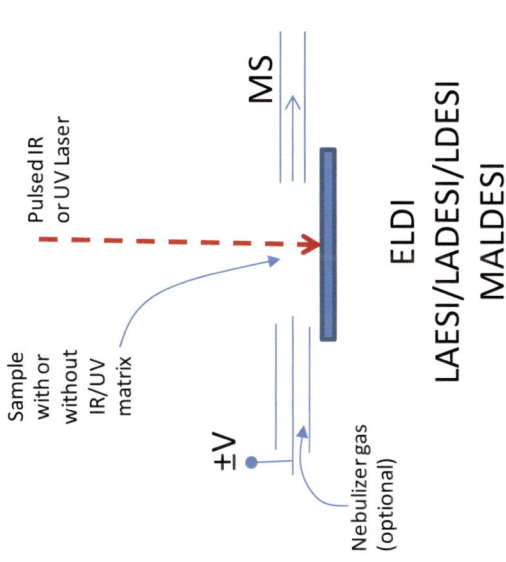

FIGURE 4-5 Laser-based ambient ionization techniques: (left) Laser Ablation-Electrospray Ionization (LAESI) and (right) Infrared Laser Ablation Metastable-induced Chemical Ionization (IR-LAMICI). SOURCE: Harris et al., 2011. Reprinted with permission. Copyright 2011 American Chemical Society.

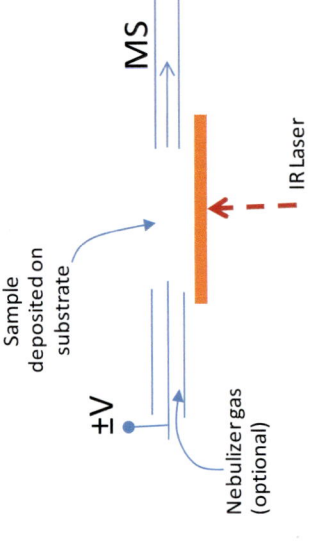

## RADIO

- Sample is aerosolized by piezo element.
- Neutral plume is entrained by electrospray droplet cloud. Analyte dissolves in droplets. Analytes in droplets ionize through ESI mechanisms.
- Sample is positioned to avoid direct interaction with electrospray liquid.

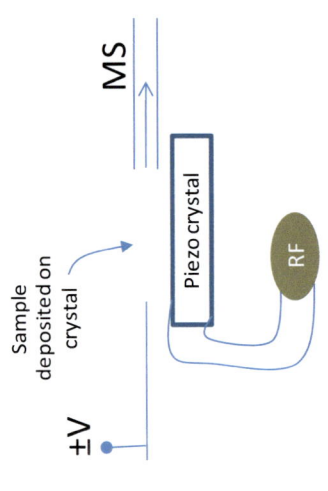

## LIAD-ESI

- Analytes are desorbed or aerosolized by laser-induced acoustic wave through special substrate.
- Neutral plume is entrained by electrospray droplet cloud. Analyte dissolves in droplets. Analytes in droplets ionize through ESI mechanisms.
- Sample is positioned to avoid direct interaction with electrospray liquid.

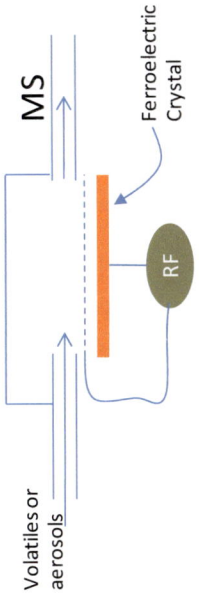

### DAPPI

- Neutral spray impinges surface.
- Spray dissolves analyte. Heated nebulizer gas enhances extraction and evaporates solvent.
- Gas-phase neutral analytes react with ionized solvent species. Ionization similar to APPI.

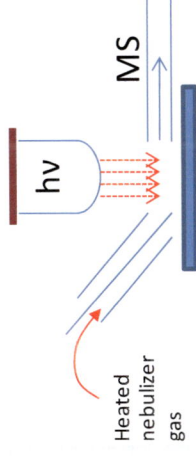

### BADCI

- Neutral spray impinges surface.
- Spray dissolves analyte. Heated nebulizer gas enhances extraction and evaporates solvent.
- Gas-phase neutral analytes react with ionized solvent species. Ionization similar to APPI.

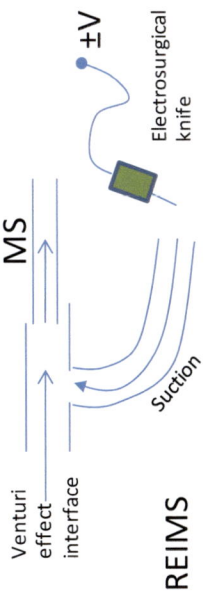

### SWIFER

- Aerosolized samples or volatiles are injected in the space above a ferroelectric crystal.
- Audiofrequency is used to rapidly switch the crystal polarization, inducing electron emission and plasma generation across the crystal-grid gap.
- Direct reaction of volatile molecules or aerosols with ferroelectric plasma species results in generation of protonated or deprotonated analyte ions.

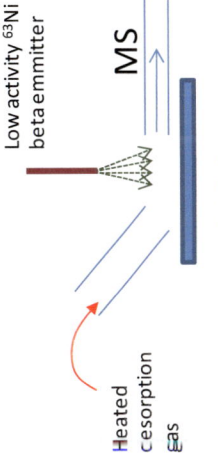

### REIMS

- Electrosurgical knife evaporates tissue by Joule heating using an electrical current. Ions are produced by chemical ionization with ionized water molecules.
- Analyte ions are transported to the mass spectrometer inlet by suction from a Venturi-effect interface

FIGURE 4-6 Schematic illustrations (this page and facing page) showing the operation of several different ion sources and sampling schemes for ambient mass spectrometry. SOURCE: Harris et al., 2011. Reprinted with permission. Copyright 2011 American Chemical Society.

The most prevalent mechanism proposed for the formation of positive ions by DART involves Penning ionization of atmospheric water molecules in collisions with electronically excited metastable helium atoms (He* 3S1, 19.8 eV; eq. 4.8) (Cody et al., 2005):

$$\text{He}^*(g) + n\text{H}_2\text{O}(g) \rightarrow \text{He}(g) + (\text{H}_2\text{O})_{n-1}\text{H}^+(g) + \text{OH}(g) + e^- \quad (4.8)$$

Protonated water clusters of different sizes react with thermally desorbed molecules, AB (g) (eq. 4.9a), by proton transfer (eq. 4.9b).

$$\text{AB}(s) \xrightarrow{\text{heat}} \text{AB}(g) \quad (4.9a)$$

$$\text{AB}(g) \xrightarrow{(\text{H}_2\text{O})_{n-1}\text{H}^+(g)} \text{AB}(\text{H}_2\text{O})_{n-2}\text{H}^+(g) + \text{H}_2\text{O}(g) \xrightarrow{-(n-2)\text{H}_2\text{O}} \text{ABH}^+(g) \quad (4.9b)$$

Dissociative proton transfer can also occur; i.e., AB does not stay intact.

Another ionization mechanism that has been observed under high grid voltages, small DART to mass spectrometer inlet spacing, and low-humidity conditions is direct Penning ionization of desorbed analytes (eq. 4.10) (Cody, 2009). This mechanism produces electron ionization-like spectra for low-polarity compounds with low proton affinities. Under these conditions, charge exchange reactions with diatomic oxygen molecular ions may also occur (eq. 4.11a and eq. 4.11b).

$$\text{He}^*(g) + \text{AB}(g) \rightarrow \text{He}(g) + \text{AB}^+(g) + e^- \quad (4.10)$$

$$\text{He}^*(g) + \text{O}_2(g) \rightarrow \text{He}(g) + \text{O}_2^+(g) + e^- \quad (4.11a)$$

$$\text{O}_2^+(g) + \text{AB}(g) \rightarrow \text{O}_2(g) + \text{AB}^+(g) \quad (4.11b)$$

Bartmess and Song et al. have thoroughly discussed the ion generation mechanisms in DART beyond those initially discussed by Cody (Song et al., 2009a, 2009b), suggesting a "transient microenvironment" mechanism. The proposed transient microenvironment mechanism hypothesizes that analyte ionization in DART occurs by proton transfer from protonated microenvironment solvent clusters rather than proceeding by proton transfer from protonated ambient water clusters. Therefore, any solvent used to dissolve the sample can have a significant effect on the ability to detect a given species depending on its proton affinity. Solvent molecular ions produced by reactions of neutral solvent molecules with metastables can also act as reagent ions, leading to both protonated analytes and analyte molecular ions. In the case of negative ion mode experiments, electron capture, dissociative electron capture, proton transfer, and anion attachment were hypothesized to be prevalent reaction pathways (Song et al., 2009a). However, as discussed above, negative ions are not expected to be important for the sensitive detection of chemical warfare agents.

The complexity of the various ionization pathways, combined with intricacies of the coupled fluid and thermal dynamics, electrical fields, and sample positioning effects present in DART ionization have been shown to lead to marked differences in the efficiency of ion transmission and hence sensitivity (Harris and Fernandez, 2009). Through simulations and experiments, it has been shown that optimum sample placement is a fine balance between (1) adequate and rapid sample heating to induce efficient thermal desorption and (2) limiting losses produced by disturbances in the ion/neutral trajectories induced by the presence of a sizeable solid sample placed within the DART ionization region. These effects, in the case of liquid or transmissive samples, can be largely mitigated by sampling in transmission mode when possible (Perez et al., 2010). In terms of ion activation, ion thermometry and computational fluid dynamics simulations have shown that DART generates ions with higher internal excitation than DESI, with a certain overlap between the two techniques depending on the operational conditions chosen. In DART, the internal energy of the analyte ion increases with increasing gas temperatures and flow rates, leading to increased in-source fragmentation within the first region of the mass spectrometer (Harris et al., 2010).

The viability of using DART for detection of chemical warfare agents was demonstrated as early as 2005 (Cody et al., 2005). This seminal DART publication describes the successful ionization and detection of members of G- and V-series nerve agents and HN-series blister agents from a variety of surfaces, including concrete. These agents were detected as their protonated molecules after a few seconds of exposure of the sample surface to the DART stream. Several recent research articles describe more extensively the application of DART to the analysis of chemical warfare agents (Nilles et al., 2009, 2010). In Nilles et al. (2009), the researchers demonstrated the detection and quantitation of GA, GB, VX, and HD in liquid samples, with recoveries better than 97 percent, linear correlation coefficients of 0.99 or better, and dynamic ranges of three orders of magnitude. Ionization chemistries were chosen such that the analyte was distributed into as few ion masses as possible. In the case of VX (Figure 4-7d), the protonated neutral species is readily, and uniquely, formed under typical DART conditions. This can be attributed to VX having a relatively high proton affinity. However, GA and GB have proton affinities close to that of ammonia (854 kJ/mol) and have a greater tendency to form the ammonium adduct. So, when analyzing these compounds, conditions were tuned to produce only $[M + NH_4]^+$ (Figure 4-7b,c). Under positive ion DART conditions, HD produces the $[M + OH]^+$ ion (Figure 4-7a). It is likely that the DART environment produces reactive species that easily oxidize HD to the sulfoxide, which has a higher proton affinity and is detected as the protonated species. Structures of the neutral molecules along with the exact masses of the detected quasimolecular ions are displayed in Figure 4-8. The production of oxidizing species is a mitigating issue in many of the plasma-based ambient ionization methods. Since factors such as relative humidity may influence the importance of oxidation chemistry (Neidholdt and Beauchamp, 2009), the viability of a detection method should be confirmed over the full range of environmental parameters in areas where it is deployed. However, should there be any doubt about the HD identification, HD can easily be confirmed with DART in negative ion mode as the chloride ($Cl^-$) adduct.

Sampling in Nilles et al. (2009) was performed by dipping the closed end of a glass capillary into a solution containing the agent of interest and a corresponding

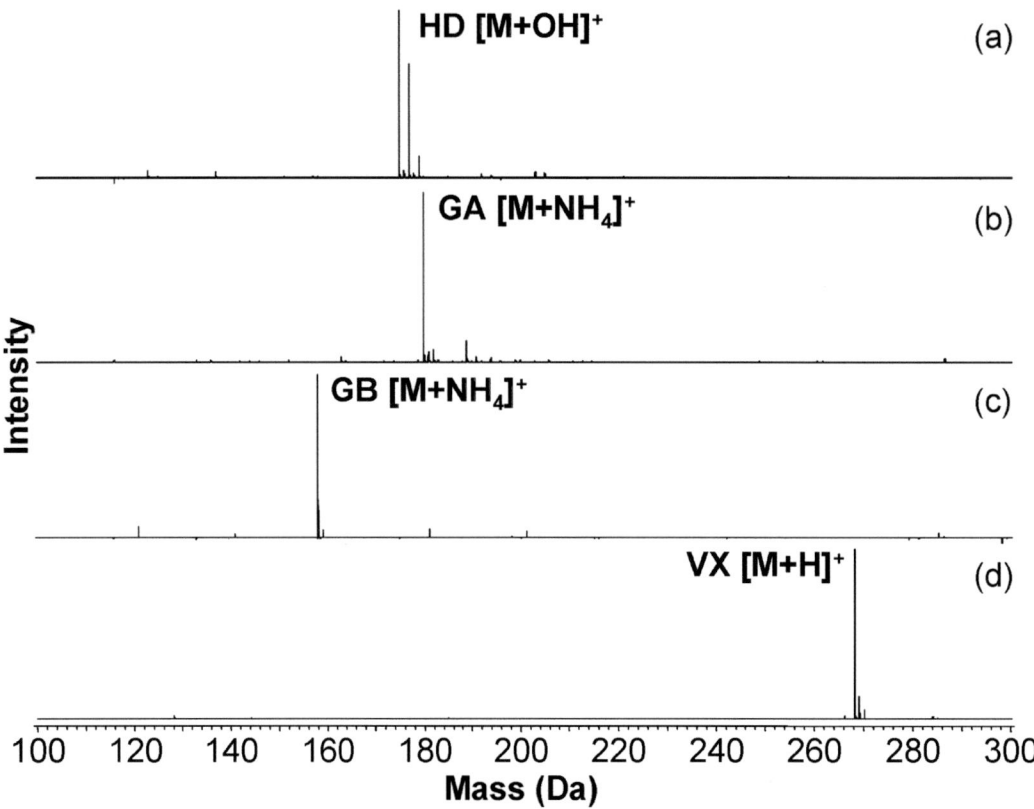

FIGURE 4-7 DART mass spectra of agent standards. (a) sulfur mustard (HD), detected as the protonated sulfoxide—the three peaks are due to the chlorine isotopes; (b) tabun (GA), detected as the ammonium ion adduct; (c) sarin (GB), detected as the ammonium ion adduct; and (d) VX, detected as the protonated species. SOURCE: Nilles et al., 2009. Reprinted with permission. Copyright 2009 American Chemical Society.

internal standard. Isotopically labeled agents $D_5$-GA, $D_7$-GB, $^{13}C_2$-HD, and $D_5$-VX were used as internal standards, and the ratio of analyte to internal standard areas used as the response variable. An expanded statistical analysis of the summary data in Nilles et al. (2009), along with additional data supplied by the authors, is provided in Chapter 5. A follow-up article by the same team demonstrated the possibility of performing rapid separations within the DART ionization region by ramping the DART gas temperature, which made it possible to distinguish the $[M + NH_4]^+$ ion of GB from isobaric analytes that could otherwise only be resolved via high-resolution mass measurements or tandem mass spectrometric experiments (Nilles et al., 2010).

The applicability of DART for detection of chemical warfare agents (CWA) on a range of real-world samples has been thoroughly evaluated (Nilles et al., 2010). Figure 4-9 illustrates high-resolution mass spectra obtained using DART for the detection of 800 ng of VX spotted on aluminum, concrete, and a bird feather. Similar to the data shown in Figure 4-7, VX is detected as the protonated molecular ion at m/z = 268.150. Aluminum and concrete yield cleaner spectra than does the bird feather, where the more complex matrix yields a wide range of detected molecules. Such species may give rise to false

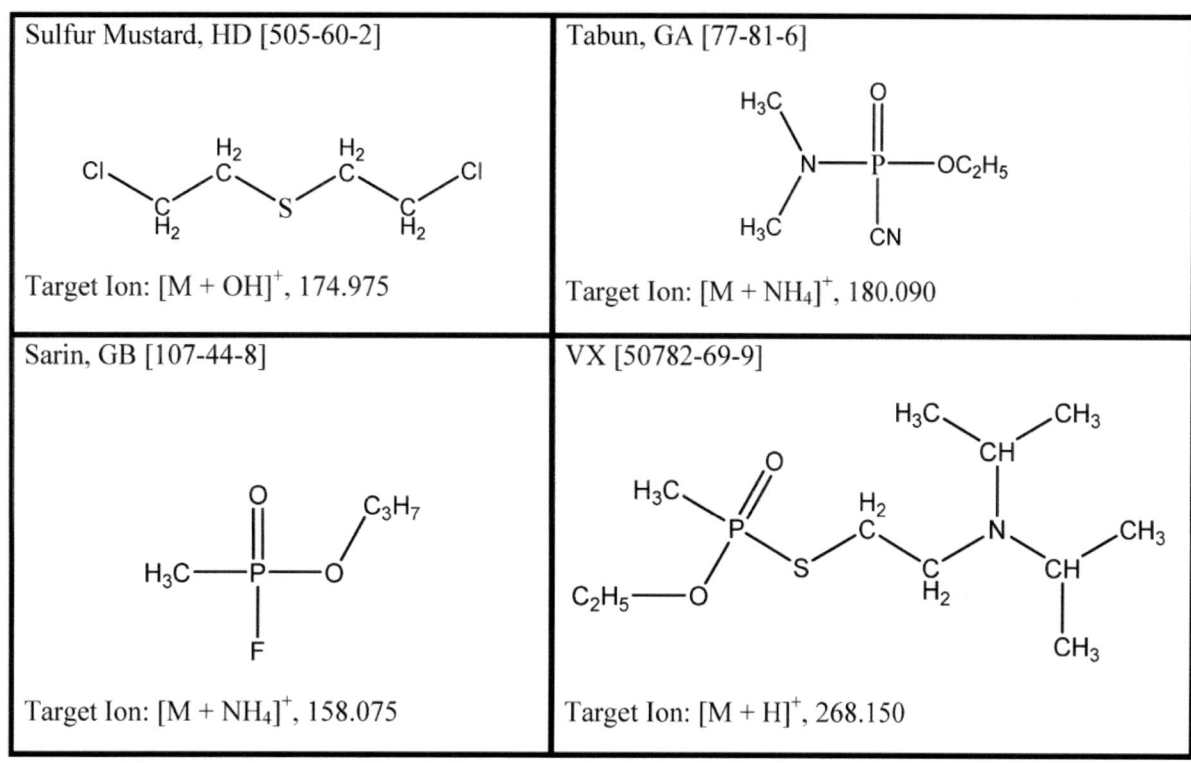

FIGURE 4-8 Structures of HD, GA, GB, and VX, with CAS designations in brackets. In each case, the target ion exact mass is given for the ions observed by DART and displayed in Figure 4-6. SOURCE: Nilles et al., 2009. Reprinted with permission. Copyright 2009 American Chemical Society.

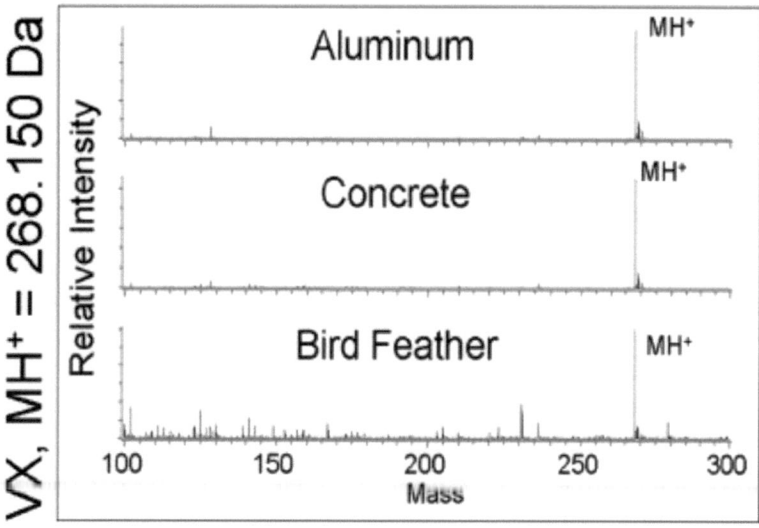

FIGURE 4-9 High-resolution mass spectra obtained by DART for 800 ng VX on aluminum, concrete, and a bird feather. VX appears as the MH+ molecular cation at m/z = 268.150 Da. SOURCE: Larame´e et al., 2008. Reprinted with permission.

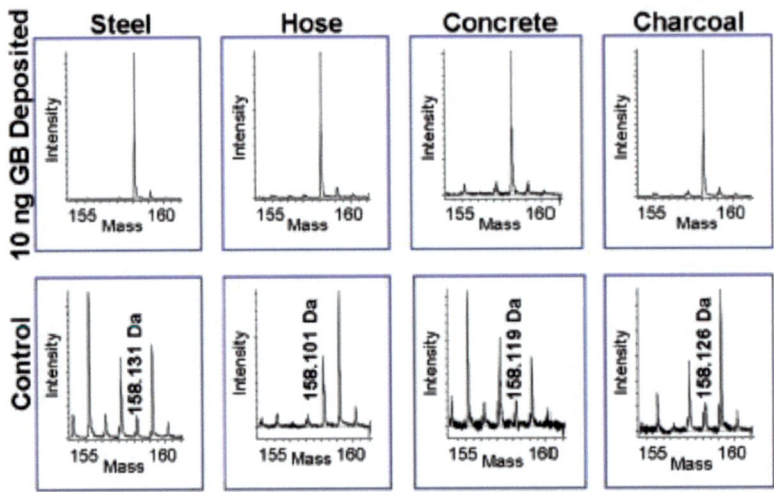

FIGURE 4-10 Surfaces of steel, rubber hose, concrete, and charcoal spiked with 10 ng GB (top row) and unspiked surfaces (bottom row). GB appears as the [MH+NH3]$^+$ cation at mass 158.075 Da. Other peaks not related to GB are seen at nominal mass = 158 but do not interfere or cause a false positive measurement because of the high-resolution mass measurement. SOURCE: Laramée et al., 2008. Reprinted with permission.

positives at the same nominal m/z of the detected CWA quasi-molecular ion. Figure 4-10 illustrates this problem when 10 ng of GB is spiked onto steel, rubber hose, concrete, and charcoal surfaces and compared to blanks run for the same substrates. In every case species are identified at the same nominal m/z = 268. Accurate mass measurement shows in each illustrates this problem when 10 ng of GB is spiked onto steel, rubber hose, concrete, and charcoal case that the major detected species in the blanks has a different elemental composition from GB. Comparison of intensities for the spiked and blank spectra suggests that detection limits at low picogram levels are readily achieved. The time-of-flight instrument used in these studies had a measured mass resolution of 6,000 (m/$\Delta$m, where $\Delta$m is the full width of the peak at half maximum) in the m/z region appropriate for the agents investigated.

## Desorption Electrospray Ionization (DESI)

DESI (Figure 4-2) was first introduced by Cooks and coworkers in 2004 (Takáts et al., 2004). DESI combines the features of ESI with those of the family of desorption ionization methods (Karas and Hillenkamp, 1988). An electrosonic spray ionization (ESSI) emitter (Takáts et al., 2005a) is used to create gas-phase solvent ions, ionic clusters, and charged microdroplets, which are directed to the surface to be sampled. An electrical potential of several kilovolts is applied to the spray solution, and pneumatic nebulization is used to assist in desolvation.

Droplet pickup has been suggested as the primary ionization mechanism in DESI, although there is evidence for chemical sputtering (reactive ion/surface collisions) and gas-phase ionization processes (e.g., charge transfer, ion-molecule reactions, and volatilization/desorption of neutrals followed by ionization) (Takáts et al., 2004; Cotte-Rodriguez et al., 2005; Takáts et al., 2005a; Costa and Cooks, 2007, 2008). According to the droplet pickup mechanism, the surface is prewetted by initial droplets (velocities in excess of 100 m/sec and diameters of less than 10 µm), forming a solvent layer that dissolves analytes on the surface. These dissolved analytes are picked up by later arriving droplets that are impacting the surface, creating secondary droplets containing the dissolved analytes. Gas-phase ions are then formed from these charged secondary droplets by ESI-like mechanisms (Takáts et al., 2005a; Costa and Cooks, 2007, 2008). The resulting gas-phase ions have low internal energies, similar to those in ESI and ESSI (Nefliu et al., 2008). The formation of cold ions gives DESI its soft ionization character that affords ESI-like spectra. DESI is able to detect species where thermal desorption is not possible, such as nucleic acids and proteins. In this respect it is very different from DART, where molecules are thermally desorbed by a hot gas stream and subsequently chemically ionized in the vapor phase.

The desorption and ionization steps occurring simultaneously in DESI can be spatially and temporally separated for further control and optimization of the experimental parameters. This is the case in a technique known as laser ablation/desorption electrospray ionization (LADESI), also referred to as electrospray-assisted laser desorption ionization (ELDI), or laser-assisted electrospray ionization (LAESI). In LADESI (Figure 4-5), a neutral plume of molecules and particles from a matrix-free sample surface are entrained with a charged solvent plume. The steps involved in LADESI ion formation are believed to encompass a desorption/ablation process induced by rapid absorption of laser energy generating a neutral plume that undergoes (1) ion/molecule reactions with ions, usually proton donors, generated by an electrospray emitter and/or (2) fusion with charged electrosprayed droplets undergoing solvent evaporation and Rayleigh discharge (Smith et al., 2002) to form ions (Huang et al., 2010).

DESI is most applicable to a solid sample or surface-bound analytes which are ionized by the charged species generated from the ESSI emitter. Recently, DESI has been extended to allow the direct analysis of liquid samples (Chipuk and Brodbelt, 2008; Ma et al., 2008; Miao and Chen, 2009; Miao et al., 2010; Zhang and Chen, 2010; Miao et al., 2011). DESI can also be coupled with solid-phase microextraction (SPME) probes to analyze gaseous species adsorbed onto the SPME fiber from the headspace above a liquid or solid surface (D'Agostino et al., 2006; D'Agostino et al., 2007; D'Agostino and Chenier, 2010).

DESI has been used for a wide variety of direct analysis, including the analysis of pharmaceutical products (Chen et al., 2005; Weston et al., 2005; Williams and Scrivens, 2005; Kauppila et al., 2006; Rodriguez-Cruz, 2006; Cotte-Rodriguez et al., 2007; Venter and Cooks, 2007), dyes on thin layer chromatography plates (Van Berkel et al., 2005), polymers (Nefliu et al., 2006), alkaloids on plant tissue (Talaty et al., 2005), pesticides spiked onto immobilized powders or various surfaces (Hagan et al., 2008), explosives on a variety of surfaces (Cotte-Rodriguez et al., 2005; Takáts et al., 2005b; Cotte-Rodriguez and Cooks, 2006; D'Agostino et al., 2006; Mulligan et al., 2006; Nefliu et al., 2006), and CWAs as well as their hydrolysis products (Cotte-Rodriguez and Cooks, 2006; D'Agostino et al., 2006; Mulligan et al., 2006; D'Agostino et al., 2007; Song and Cooks, 2007; D'Agostino and Chenier, 2010). The analytical capabilities of DESI-MS are illustrated by an ability to distinguish diseased and nondiseased tissue based on their chemical signatures (Wiseman et al., 2005). In these studies, qualitative changes in the lipid profiles were obtained that distinguished the tumor from the nontumor region of a biopsied liver adenocarcinoma tissue.

Analysis of CWAs by DESI has been reported using two different approaches, involving the use of SPME fibers (D'Agostino et al., 2006; D'Agostino et al., 2007; D'Agostino and Chenier, 2010) and direct sampling (Cotte-Rodriguez and Cooks, 2006; Mulligan et al., 2006; Cotte-Rodriguez et al., 2007; Song and Cooks, 2007). The first report of an analysis of CWAs by DESI appeared in early 2006. It reported the detection of the nerve agent simulant triethyl phosphate as well as two nerve agents, GB and soman. The acquired DESI-MS/MS data for CWAs was similar to that obtained by traditional liquid chromatography ESI-MS/MS. SPME headspace sampling followed by the direct DESI-MS analysis took less time for sample preparation and analysis than did aqueous extraction and liquid chromatography ESI-MS/MS analysis, indicating that DESI could be a good candidate for high-throughput analysis (D'Agostino et al., 2006). Later, more complex CWA samples were analyzed by SPME DESI-MS/MS, with detection limits in the sub-nanogram range (D'Agostino et al., 2007; Song and Cooks, 2007). Very recently, ion mobility separations were used as a part of rapid SPME DESI-MS confirmation method for some common organophosphorus CWAs (D'Agostino and Chenier, 2010). In terms of direct sampling of CWAs, the detection specificity can be significantly enhanced by "reactive" DESI. In Song and Cooks (2007), this mode of DESI analysis involved the addition of boric acid to the spray solvent to enhance the analysis of the hydrolysis products of phosphonate esters through heterogeneous ion/molecule reactions. On average, sub-nanogram detection limits were obtained for methylphosphonic acid (MPA), ethylphosphonic acid (EMPA), and isopropyl methylphosphonic acid (IMPA) from complex matrices (Song and Cooks, 2007). When a methanolic solution of these compounds was deposited on a clean Teflon surface, 10 pg each of EMPA, IMPA, and MPA could be readily detected (Song and Cooks, 2007).

When charged droplets impinge on a surface from an electrospray source driven by a flow of sheath air, they pick up material and will eventually remove soluble deposits from the surface. Organic material on the surface is removed by impinging charged droplets that rebound and repel each other, developing into a spreading plume that can distribute desorbed material over a wide area. Specially designed electrospray sources have been developed to clean surfaces for high-value end products such as silicon wafers. It would not be desirable to disperse localized CWA deposits over a larger area, leading to more widespread contamination, which might include the detection instrumentation. Due to their exceptionally low flow rates, use of nanospray sources could help to alleviate the spread of sampled species from the surfaces (Roach et al., 2010).

### Nonproximate Analysis by Ambient Mass Spectrometry

The need for field-portable nonproximate detection is apparent for compounds harmful to health or the environment, such as toxic industrial species, explosives, CWA and environmental toxins (Badman and Cooks, 2000; Gao et al., 2006; Mulligan et al., 2006; Chaudhary et al., 2006). Remote detection of these classes of compounds increases safety and is more convenient and time responsive than the laboratory proximate analysis of field samples. Several systems providing this capability have become commercially available. For example, the Bruker Scanning Infrared Gas Imaging System 2 is a sensing system for remote infrared detection of hazardous gases in industrial, environmental, and homeland security applications.[2] In traditional MS analysis of field samples with ambient mass spectrometry techniques, the ionization source is located close to the atmospheric pressure inlet of the mass spectrometer, limiting the analysis to small, well-defined sample substrates (Dixon et al., 2007). A highly desirable feature to enhance the utility of ambient mass spectrometry is implementation of a system having capability for nonproximate analysis, whereby the ion source is remotely coupled to the MS detector.

DESI has been used for the direct and nonproximate detection of trace amounts of explosives (RDX, HMX, PETN, TNT, Composition C4, and TATP) and the nerve agent simulant dimethyl methylphosphonate (DMMP) from ambient surfaces up to 3 m from the mass spectrometer using a stainless steel transfer tube (Figure 4-11a) (Cotte-Rodriguez and Cooks, 2006). In that study, the analyte of interest was desorbed from the surface and ionized by charged droplets generated using a DESI spray source. The resulting droplet plume, along with entrained analyte ions, was transported for analysis through the stainless steel transfer tube with the help of the vacuum of the mass spectrometer. Using this apparatus, the limits of detection for explosives RDX, HMX, PETN, TNT, Composition C4, and TATP and of the CWA simulant DMMP were in the low-nanogram range with a 3-m transfer tube. With a shorter 1-m transfer tube, limits of detection for RDX and DMMP were in the subnanogram range (Cotte-Rodriguez and Cooks, 2006). The same research group later expanded this experimental methodology for nonproximate analysis by performing reactive DESI to increase both selectivity and sensitivity for the detection of target analytes with proton affinities above a certain value. For example, in the analysis of 2,4,6-triphenylpyridine, ammonia vapor was introduced into the ion transfer tube to remove abundant ions (chemical noise) arising from neutral species with proton affinities lower than ammonia (Figure 4-11b) (Cotte-Rodriguez et al., 2007). In addition, by optimizing the temperature of the transfer tube, the fragmentation and the formation of stable adduct ions could be controlled (Cotte-Rodriguez et al., 2007).

Ion transfer in nonproximate analysis can also be assisted by controlling airflow to enhance transport of ions to the atmospheric sampling interface of the mass spectrometer. Implementations include the use of a commercial air ejector (Dixon et al., 2007) (Figure 4-12 [top]) and airflow-assisted ionization (AFAI) (He, 2011) (Figure 4-12 [bottom]). As shown in Figure 4-12 (top), the commercial air ejector is used to transport ions formed either by ESI or by DESI from a surface, from the point of formation to the MS inlet and then through a flexible polyethylene tube to the MS. In the AFAI method, high-flow-rate airflow extraction effectively captures the charged droplets containing analytes from the DESI desorption site, which are then transferred 0.5 m to the inlet of the MS.

Analytes in a remote location can also be transferred as neutral species for MS detection. Either the remote analyte sampling, transport, and ionization relay (RASTIR) method (Dixon et al., 2008) (Figure 4-13 [top]) or neutral desorption-extractive

---

[2]Additional information is available at http://www.brukeroptics.com/gas_detection.html.

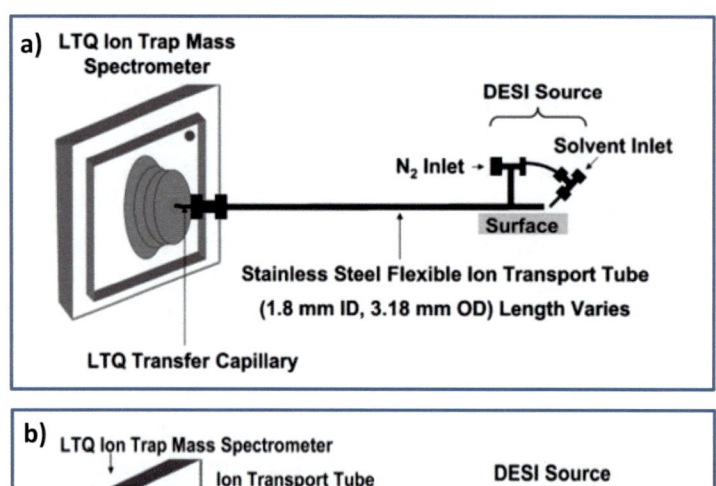

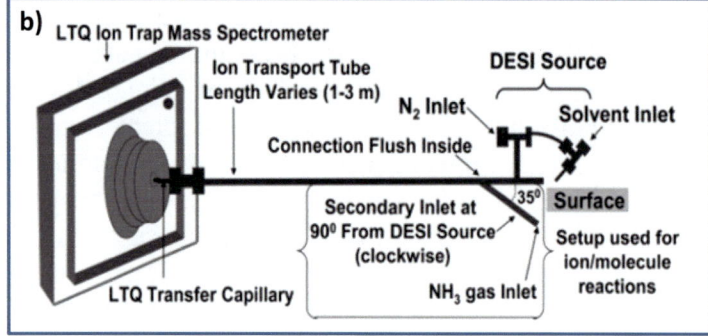

FIGURE 4-11 DESI remote sampling techniques. (a) Experimental arrangement for non-proximate analysis using ambient mass spectrometry. The DESI source is remotely located from the mass spectrometer, with ions transported over distances up to 3 m through a flexible stainless steel tube. SOURCE: Cotte-Rodriguez and Cooks, 2006. Reproduced with permission of the Royal Society of Chemistry. (b) Modified experimental arrangement for implementation of reactive DESI in the transfer tube, in which a secondary inlet allows the introduction of a neutral reagent gas. Ammonia, for example, scavenges protons from trace species with lower proton affinity to remove chemical noise from the spectra. SOURCE: Cotte-Rodriguez et al., 2007. Reprinted with permission. Copyright 2007 American Chemical Society.

electrospray ionization (ND-EESI) (Chen et al., 2009) (Figure 4-13 [bottom]) can be used to implement this approach to nonproximate analysis. Based on the venturi effect, the RASTIR source creates an induced vacuum from high-pressure nitrogen gas flow, which can be used to transfer low-volatility analytes for subsequent ionization by ESI. In the ND-EESI experiment, a trace amount of explosives from biological surfaces such as human skin can be detected rapidly through a long transfer tube (up to 4 m) after interacting with a charged droplet beam generated by ESSI. In the ND-EESI experiment, selective ion/molecule reactions can also be implemented, which further enhances the detection selectivity and the sensitivity.

In order to scan large surface areas for trace chemical detection, modified DESI has been introduced, including multiple sprayer DESI (Cotte-Rodriguez et al., 2007) (Figure 4-14 [top]) and large-area sprayer enclosure DESI (Figure 4-14 [bottom]) (Soparawalla et al., 2009). Although DESI can be used for nonproximate analysis of individual samples of interest within a small surface area, the total analysis time required will be substantial if large objects are to be examined. A multiple DESI sprayer setup (using three sprayers covering areas of about 70 $mm^2$) (Figure 4-14 [top]) was designed to address this issue, and it could also be used for high-throughput analysis. An

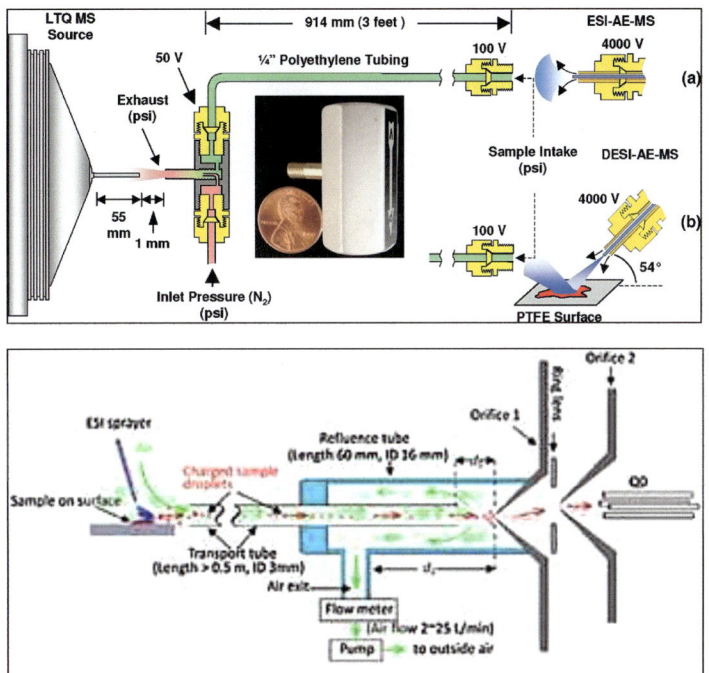

FIGURE 4-12 DESI remote sampling techniques using AE and AFAI. (Top) Schematic of the commercial air ejector (AE) interfaced to a linear quadrople ion trap-MS with an ESI or DESI ion source; the inset photograph shows the actual injector. SOURCE: Dixon et al., 2007. Copyright 2007. Reprinted with permission from Elsevier. (Bottom) Schematic showing operation of the non-proximate airflow-assisted ionizer (AFAI) SOURCE: He et al., 2011. Copyright 2011 John Wiley & Sons. Reprinted with permission.

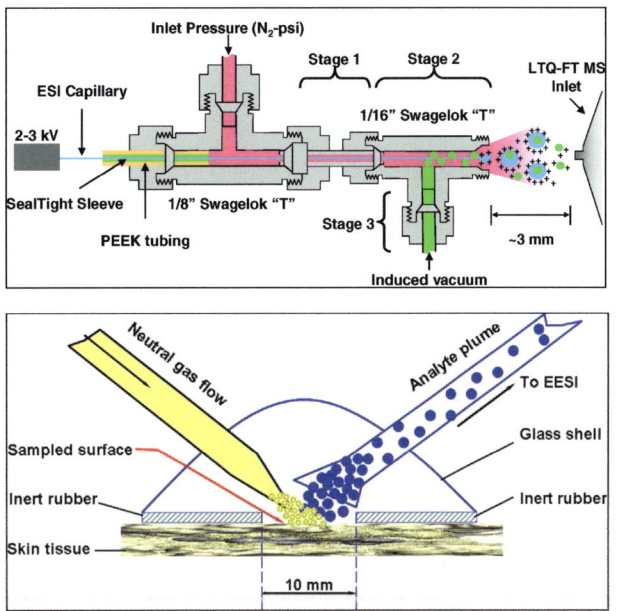

FIGURE 4-13 RASTIR and ND-EESI schematics. (Top) Labeled schematic of a RASTIR assembly showing the high-pressure region (stage 1), high-velocity jet/entrainment/exhaust region (stage 2), and the vacuum region (stage 3). SOURCE: Dixon et al., 2008. Reprinted with permission. Copyright 2008 American Chemical Society. (Bottom) Schematic of an air-tight neutral desorption enclosure for the ND-EESI method. SOURCE: Chen et al., 2009. Copyright 2009. Reprinted with permission from Elsevier.

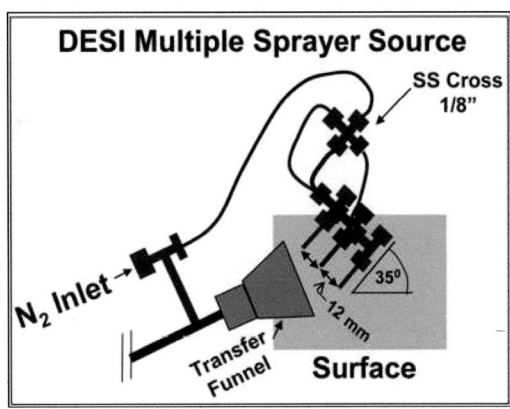

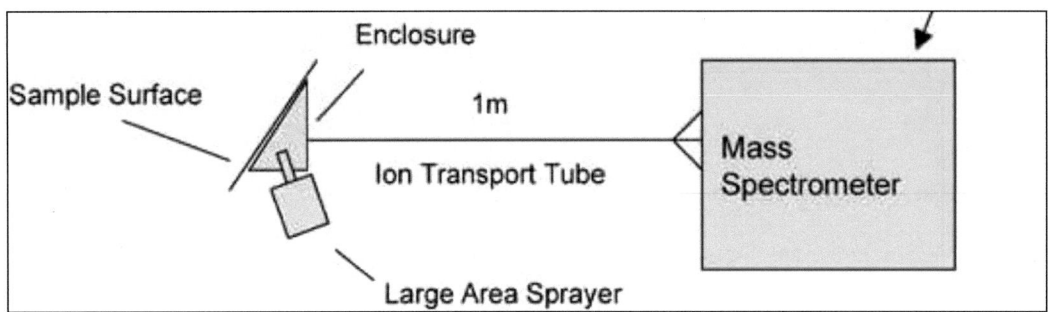

FIGURE 4-14 Multiple-sprayer and nonproximate large-area sprayer DESI setups. (Top) Multiple-sprayer DESI setup used for high-surface-area screening as well as emitters that are welded together. SOURCE: Cotte-Rodriguez et al., 2007. Reprinted with permission. Copyright 2007 American Chemical Society. (Bottom) Nonproximate large-area DESI setup consisting of a large-area sprayer coupled to an ion transport tube. SOURCE: Soparawalla et al., 2009. Copyright 2009 John Wiley & Sons, Inc. Reprinted with permission.

alternative large-area analysis DESI source is shown in Figure 4-14 [bottom]. This system uses two coaxial stainless steel tubes, with the internal tube having an inner diameter of 1 cm. High-pressure nitrogen gas (300 psi) and solvent were propelled through the annular space, covering a large-area (up to 3 $cm^2$ with effective ionization) screening surface, which was interlocked with the transfer tube via a plastic enclosure (Soparawalla et al., 2009). These large-area nonproximate analysis modes, which have been particularly well implemented with DESI, could find uses in mapping chemical agent contamination on equipment or building surfaces.

Preliminary reports on the remote tethering of a handheld DART ion source to a mass spectrometer through a 1.5-m-long transfer line have recently been presented (Musselman, 2011), but extensive testing has not yet been reported. Owing to its inherent simplicity, robustness, and suitability for small molecule detection, remote-sampling DART has the potential for becoming a key means for surface sampling and analysis under harsh field conditions.

Most of the ACWA scenarios for ambient ionization MS suggested in this report will require a sampling wand. However, the degree of target species dispersion caused by these wand based-based sampling systems has not been described in the literature, and probably has not been characterized. The suspected proclivity of DESI for dispersing target species more extensively than DART is based on their respective fundamental operational properties and the basic physics of each technique, and not on data for specific wand designs (which are not available).

Overall, DART, DESI, and other nonproximate sampling and analysis approaches have expanded the capability of mass spectrometry in analyzing dangerous compounds with improved safety for examiners. In addition, portable (Mulligan et al., 2006) or miniaturized (Gao et al., 2006) mass spectrometers could be coupled to the variety of the remote sampling devices and operated at distance for increased flexibility during in situ analysis (Cotte-Rodriguez et al., 2007).

## POTENTIAL ROLES FOR AMBIENT MASS SPECTROMETRY IN THE ACWA PROGRAM

Virtually all early applications of ambient mass spectrometry were based on thermal desorption combined with atmospheric-pressure chemical ionization (APCI) (Van Berkel et al., 2008). Pioneering studies by Horning and coworkers in the 1970s employed samples collected on platinum wires, which were desorbed using a heated gas flow and ionized by APCI (Dzidic et al., 1975). A variety of commercial systems were developed based on ambient mass spectrometry, mainly to detect explosives and drugs. Nonproximate analysis was accomplished using a heated surface sampler coupled to a mass spectrometer through a 10-m heated transfer line (Stott et al., 1993). These investigations promoted the concept of near-real-time analysis of target species without need for sample preparation. With complex mixtures, selected reaction monitoring using MS/MS analysis performed using triple quadrupole mass spectrometers provided confident identifications and avoided false positives. Of particular note is an early study by Fite and coworkers at Extrel Corporation, supported by the U.S. Army Program Manager for Chemical Demilitarization, in which an atmospheric pressure ionization tandem quadrupole mass spectrometer system was used to directly detect GB and VX in air with a 15-sec response time and with detection limits of 7.2 ppt for GB and 6 ppt for VX (Ketkar et al., 1991b). A corona discharge provided protonated water reagent ions. That selected reaction monitoring could be used to ensure positive detections was also demonstrated, but the detection limit was reduced for GB to 14.1 ppt and for VX (cleaved to its G-analog) to 100 ppt. The same group provided an insightful analysis of the influence of coexisting analytes with higher proton affinities than the target molecules (Ketkar, 1991a).

This evolutionary growth of ambient mass spectrometry spiked beginning with the development of DESI, whose first publication was in November 2004, and DART, whose first publication was in March 2005. The commercialization of these experimental methodologies and the many variations they inspired have led to widespread applications of ambient mass spectrometry. While DESI in many ways is unique, it is of interest to note that DART is a variation of thermal desorption combined with APCI. The design of the DART source is unique, however, in that it isolates the corona discharge from the target sample, filters out directly formed ions from the discharge, and relies on excited neutral species to generate reactant species that subsequently chemically ionize species present in the vapor phase or thermally desorbed from surfaces by a variable temperature stream of heated gas. This avoids having target molecules interact with energetic electrons and ions produced in the corona discharge. The use of DART's gas phase desorption/ionization process is also logistically less complex than DESI's liquid jet-based splatter desorption and ionization process.

As detailed above, ambient ionization mass spectrometry technology has made great strides since its introduction in 2004/2005. A range of systems and system components are now commercially available from a variety of U.S.-based analytical instrument companies. A number of these companies and their major commercial offerings are listed in Appendix C.

Table 4-3 provides a detailed summary of the capabilities and limitations of ambient mass spectrometry (DART and DESI) compared to the existing vapor monitoring technology and measurement strategies (DAAMS and MINICAMS). Table 4-4 summarizes some of the relevant strengths and weaknesses of DESI and DART for the chemical agent contamination characterization scenarios described in this report. Both DESI and DART have demonstrated an impressive capability to detect trace levels of ACWA-relevant chemical agents in solution and adsorbed on surfaces. DART can also be modified to detect trace levels of airborne agent in real time. Finally, because most of the published applications of both DESI and DART, including the chemical warfare agent and agent simulant studies described above, are laboratory-scale experiments performed under carefully controlled conditions, in which very small amounts of the target species are present on sample substrates of modest extent, the mobilization and dispersion of target species were seldom either a scientific or a safety issue and were not generally considered or commented on in those studies. However, the potential ACWA applications defined in this report may involve much higher levels of chemical agent contamination where target species dispersion is potentially much more serious. That is why target species dispersal is potentially much more important for envisioned ACWA plant applications than in much smaller scale laboratory demonstrations.

These summaries, along with the discussions of ambient mass spectrometry relevant to the detection of chemical warfare agents presented in this chapter, form the basis for the committee's findings and recommendations offered below. If the Assembled Chemical Weapons Alternatives managers decide that the potential advantages of ambient ionization technology applied in any of the scenarios developed in Chapter 3 and discussed further in Chapter 5 are worth the cost and effort required to acquire and integrate new detection technology, the following findings and recommendations should help guide specification, acquisition, and integration activities.

TABLE 4-3 Capabilities and Limitations of Ambient Mass Spectrometry (DART and DESI) and Existing Vapor Monitoring (DAAMS and MINICAMS) Measurement Strategies

| Ambient Mass Spectrometry (DART and DESI) | Vapor Monitoring (DAAMS and MINICAMS) |
|---|---|
| **Capabilities** | |
| <ul><li>Direct measurements from vapor (DART), liquid (DART and DESI), or solid (DART and DESI) matrices</li><li>Possibility of surface spatial discrimination with user-defined resolution (ca. 50 μm–several centimeters, only DESI)</li><li>Ability to track agent concentration gradients in the vapor phase (only DART)</li><li>Ability to remotely sample through wands</li><li>Simultaneous measurement of multiple agents during changeover and decommissioning with high selectivity through analytical and confirmatory signals</li><li>Potential for rapid changeover of multiple types of ion sources for enhanced selectivity and spatial resolution</li><li>Potential for validation through monitoring both agent ions and product degradation fragmentation ions</li><li>Analysis in near real time (msec or sec)</li></ul> | <ul><li>Demonstrated reliability for bulk vapor analyses in chemical demilitarization plants</li><li>Established STEL values based on toxicological data that correspond directly with action protocols</li><li>False positive rates reasonably well characterized through long record of utilization</li><li>Agent calibration reference standards are readily available and utilized with regular frequency</li><li>Established regulatory guidelines are available for vapor and vapor headspace analyses</li><li>Potentially easier instrumental clean-up from contamination by analyzing concentrated samples</li></ul> |
| **Limitations** | |
| <ul><li>Potential for instrumental contamination owing to high sensitivity</li><li>Lack of standard reference materials for agents on various surfaces</li><li>Limited knowledge regarding correspondence of surface concentrations with those in the gas-phase</li><li>Lack of regulatory guidelines for concentrations of agent on surfaces</li><li>Expensive consumables: solvents, high-purity gases</li></ul> | <ul><li>Primarily utilized for vapor analysis (limited for liquids and solids)</li><li>Indirect analysis of liquid and solid samples by vapor measurements (e.g., via swabs, etc.)</li><li>Significant time and exposure is dedicated to changeover activities (e.g., changing detectors from FID to halogen specific detection, etc.)</li><li>Typically process detection is set to monitor a single agent</li><li>Relatively slow analysis rates (ca. 5 min for MINICAMS)</li></ul> |

TABLE 4-4 Comparative Capabilities and Limitations of DART and DESI for Characterization of Contamination by ACWA-Relevant Chemical Agents (GB, VX, HD)

| Capability/Limitation | DART | DESI |
|---|---|---|
| Real-time mapping of surface adsorbed agent distributions | Yes – with moderate spatial resolution (~1 cm) | Yes – with high spatial resolution (~50 µm to 1 cm) |
| Real-time analysis of liquid solvent agent concentrations | Yes – with appropriate liquid-phase reference standards | Yes – with appropriate liquid-phase reference standards |
| Near-real-time vapor-phase agent concentration measurements | Yes – with direct sampling | Yes – by bubbling vapor through solvent and liquid-phase analysis or direct analysis of SPME adsorbed sample |
| Real-time vapor-phase agent concentration measurements and airborne plume tracking | Yes – with direct sampling | No – requires liquid- or solid-phase sample collection step |
| Non-proximate liquid pool or contaminated surface measurements | Yes – but few demonstrated wand configurations | Yes – a range of demonstrated wand configurations |
| Ionization source logistics | Gas phase – relatively simple to deploy and control | Liquid phase – moderately more difficult to deploy and control |
| Agent dispersion during measurement | Moderate issue – can control agent vaporization rates and atmospheric pressure gas-phase diffusion is slow | Moderate issue – liquid jet can be miniaturized, and dispersed agent can be intercepted by shields |

## FINDINGS AND RECOMMENDATIONS

**Finding 4-1.** When compared to existing vapor monitoring (DAAMS and MINICAMS) measurement strategies, ambient ionization mass spectrometry provides the following capabilities for detection and quantitation of chemical agents and their degradation products (see also Table 4-3):

- Exceptional sensitivity and selectivity.
- Direct measurements from vapor, liquid, or solid samples.
- Real-time measurement with minimum response times of milliseconds to seconds.
- Multiagent detection capability and degradation product monitoring.
- Measurements of variations in spatial and temporal concentration.

**Finding 4-2.** DART, DESI, and other emergent ambient ion sources can be considered for surface analysis. A range of different technologies can be employed for liquid and ambient vapor ionization. Both DESI and DART are able to sample and identify molecular species on surfaces in real time (less than 5 sec per measurement).

**Finding 4-3.** DESI requires a solvent, usually an organic or organic/water mixture, with a few percent acid to enhance protonation of the target molecules. A high-velocity gas flow concentric with the electrospray tip directs charged droplets toward the sample. This plume can easily scatter liquid and solid material. Further dispersal of chemical agent that might lead to contamination of adjacent surfaces, especially the mass spectrometer, is not a desirable result.

**Finding 4-4.** DART, using a controlled flow of heated gas typically comprising helium or a nitrogen/helium mixture, provides the capability to heat, evaporate, and subsequently ionize the target species with modest target agent dispersal. Furthermore, it is possible to vary the temperature of the flowing gas to carry out temperature programmed desorption, observing in turn species with a wide range of volatilities. In addition to surface sampling capabilities, the DART source is an efficient atmospheric pressure chemical ionization source and hence can ionize and detect species at low concentrations in the vapor phase.

**Recommendation 4-1.** While both DESI and DART have been demonstrated to have excellent sensitivity for detecting chemical agents in liquids and on a wide range of surfaces, if only one technique is adopted, DART is preferable for the potential ACWA utilization scenarios, based on its lower dispersion of target species, utilization of a gas rather than a liquid as the working medium, and ability to efficiently ionize and detect trace levels of species in the gas phase. A DESI system with a cover shield to intercept dispersed contaminants may also be applicable. The use of instrument shields to minimize agent contamination would have to be investigated during instrument test and evaluation activities.

**Finding 4-5.** The platform configuration most likely to satisfy the analytical needs put forward in the various scenarios in Chapter 3 for different waste streams consists of a cart-mounted or handheld mass spectrometer equipped with a modified interface to accommodate a special remote sampling wand, a surface ambient ionization source combined with a vapor ambient ionization source, and any sampling accessories. Ambient ionization mass spectrometry systems backed by an uninterrupted power supply will allow portability between different rooms or site areas without breaking vacuum. Careful attention to instrument shielding and sampling wand design and implementation can reduce the possibility of agent contamination during instrument use.

**Finding 4-6.** At the highest levels of sensitivity, detection of chemical agents requires high mass resolution to distinguish the agents from isobaric trace species with the same nominal mass. A minimum resolution of 10,000 ($m/\Delta m$, where $\Delta m$ is the full width at half maximum) is required for this purpose. This value can be accomplished with an orthogonal sampling time-of-flight (TOF) mass spectrometer incorporating a reflectron. High-end instruments of this type can provide mass resolution 1.5-5 times this value and would be preferable if not for their cost and size. Selective reaction monitoring, in which an ion structure is confirmed by specific fragmentation pathways, can avoid false positives for chemical agent identification without requiring high mass resolution. While this can be accomplished with newer TOF-TOF instruments, the suggested mass analyzers are either a triple quadrupole or a linear ion trap.

**Recommendation 4-2.** The Army should specify a list of requirements for an ambient ionization mass spectrometer system that would implement analytical capabilities specifically designed to respond to the challenges summarized in the different scenarios described in Chapter 3. Suggested mass analyzers are either triple quadrupole or linear ion trap, because both can be operated in selective reaction monitoring modes for validation of agent identification and reduction of false positives without requiring high mass resolution. The instrument's atmospheric pressure interface should be fitted with an optionally heated transfer line designed to serve as a multifunction sampling wand. Ionization can occur either at the end of the wand (desorbed surface species or ambient vapor) or at the atmospheric pressure sampling orifice of the mass spectrometer (vapor). In the first case, ions are transported, whereas in the second case, neutrals are generated and then transported to the ionizer. A system employing a wand should be tested for the efficacy of the analytical methodology to trace an expanding vapor plume back to its source. This would be especially beneficial in identifying and locating Type I and possibly Type II occluded spaces (see Box 3-3) as well as in identifying leaker munitions in both storage and processing areas. The overall ambient ionization mass spectrometric system should be portable, either cart-mounted or handheld, for maximum utility.

**Recommendation 4-3.** The sampling wand should accommodate a variety of sampling modes and interchangeable ion sources:

- A compact DART (and possibly DESI) source that can be mounted either at the end of the wand, with ions sampled through the wand, or directly on the mass spectrometer, with neutrals sampled through the wand.
- A thermal neutral desorption mode where hot gas is blown toward the surface and the desorbed neutrals are collected by a suction interface that directs them toward the ionization source attached to the mass spectrometer.
- A gas sampling mode where ambient air is drawn into the interface but no gas is blown to any surfaces.
- A sampling port that allows the user to manually wipe an area and either place the wipe directly on the sampling wand or insert it through the port into the plasma of the DART source attached to the mass spectrometer.
- A high-sensitivity air monitoring mode where a solid-phase-microextraction fiber is exposed to the vapor to be sampled and directed into the plasma source.
- A high-sensitivity surface sampling mode where a polydimethylsiloxane membrane is attached to the surface to be sampled for an extended period of time and then exposed directly to the plasma for desorption ionization.
- A liquid sampling mode that allows manual sampling of a liquid pool or drip.

**Finding 4-7.** The chemical agents VX, GB, and HD have markedly different acid-base properties in the gas phase. This is reflected in the quasimolecular ions observed and monitored when DART is employed for their detection. Only VX is observed as the protonated molecular ion, GB is detected as an adduct with ammonium ion, and HD is detected as the oxidized sulfoxide derivative, which has a proton affinity significantly higher than that of the parent molecule. While it has been demonstrated that all three species can be detected with high sensitivity in laboratory studies using the same instrumental conditions, it is not obvious that this multiagent detection capability will be possible when this experimental methodology is deployed and used in working environments. For example, ammonia in laboratories can originate from human breath and would likely not be as abundant in Class A environments where workers are in DPE suits and might have to be added to the sampling flow. In addition, the extent of sample oxidation in ambient ionization sources is known to be dependent on source operating conditions and environmental factors such as the relative humidity.

**Finding 4-8.** MINICAMS are designed to monitor relatively large areas and alarm when levels exceed a specified limit. Although regarded as near-real-time detectors, they typically require 5 to 10 min for a single analysis. This is not particularly efficient when attempting to locate and define an area of agent contamination. For example, when exposure to agent is possible, MINICAMS sampling ports are moved over worker DPE suits to see if agent levels exceed 1 vapor screening level, pausing in four quadrants to take a measurement. This is a time-consuming procedure.

**Recommendation 4-4.** Procedures developed and optimized in laboratory environments for the real-time detection of chemical agents using ambient ionization mass spectrometry should be verified in all working environments where they are likely to be deployed, using actual sample materials (e.g., activated charcoal from filter beds and worker masks, DPE suit material, and polymer-coated concrete).

**Recommendation 4-5.** Procedures should be developed for using ambient ionization mass spectrometry (e.g., DART, large-area DESI) to check worker DPE suits for contamination when workers are exiting Class A work areas. This approach could greatly reduce the time required for this activity, including verification of the effectiveness of decontamination procedures carried out prior to DPE suit removal when agent is detected.

**Finding 4-9.** Mass spectrometric detection methodology is able to provide relative quantification of analyte species, including chemical agents, over several orders of magnitude with excellent linearity between concentration and response for both gas- and liquid-phase analytes with appropriate sampling and ionization methods. Absolute quantification can be provided with appropriate reference standards. This is true of the ambient ionization mass spectrometric methods as well. However, quantifying amounts of target species on surfaces and adsorbed in solid substrates is more problematical.

**Recommendation 4-6.** The Army should develop reference standards to permit calibration of mass spectrometric instruments using DART (or other deployed ambient ionization sources) for analysis of chemical agents in gases and liquids. In the case of gas-phase samples, it would be useful to develop a reference standard that reliably provides a vapor-phase concentration equal to 1 vapor screening level of the target agent, both to quantify measurements and to verify acceptable performance during critical operations. Even though they may be less reliable for quantitative analysis, calibration standards and procedures should also be developed that ensure acceptable sensitivity for detection of trace amounts of agents on relevant surfaces.

# 5
# Statistical Methods and Measurement

## OVERVIEW

The central focus of the present committee's activities is to evaluate the potential for new measurement technologies to make real-time and localized measurements for the presence of chemical agents at the Pueblo Chemical Agent Destruction Pilot Plant (PCAPP) and the Blue Grass Chemical Agent Destruction Pilot Plant (BGCAPP), including the possibility of making quick measurements on surfaces and detecting and quantifying low-level contamination.

New measurement capabilities might make possible more efficient and more reliable procedures. For example, using some of the new analytical technologies identified in Chapter 4, there are now quick and reliable means to detect and, if appropriate calibration standards are available, to quantify agent contamination on surfaces and hidden in crevices or other occluded places in machinery or building materials. Using these methods, it is now possible to more efficiently identify and delimit local hot spots of contamination, allowing a more efficient sorting of waste streams from the deconstruction process into segregated "contaminated" and "not contaminated" streams, each of which could get appropriate handling. The ability to conduct such segregation could enhance the speed of handling contaminated material and free the handling of uncontaminated material from onerous precautionary processes that may be unecessary in the absence of contamination.

To begin, some broad and overarching observations are worth noting. First, there are two broad categories of detection and measurement issues: (1) "monitoring," which is used to check the adequacy of and detect inadvertent failures of current work practices, procedures, and protocols; and (2) "detection and characterization of contamination," which aims at providing the basis for confident segregation and characterization of waste streams, identification of contaminated equipment, or parts of equipment, or areas of the facility that might have become contaminated with agent through routine operations or mishaps and malfunctions.

Overall, the following monitoring and measurement issues arise:

- *Safety of workplace air for workers.* Airborne agent monitoring is routine at chemical weapons demilitarization facilities, both to assure short-term air standards to protect against acute agent exposure effects and to establish that longer term standards designed to protect against any toxicity from ongoing lower-level exposure have been met. The instruments and methods used for these purposes were reviewed in earlier National Research Council (NRC) reports (e.g., NRC, 2005a) and are not the subject of this report. Proactive evaluations of potential agent vapor sources that might lead to contaminated air (e.g., identifying potential agent vapor sources during maintenance or deconstruction activities) are a second element that may be aided by the new technologies discussed in Chapter 4.

- *Contamination of local ambient air.* During operations, the heating, ventilation, and air conditioning (HVAC) outflow is monitored with traditional methods for measuring airborne agents. During deconstruction, the containment and processing of the workplace air will no longer be in operation, and fugitive agent emissions from the site resulting from off-gassing from no-longer-contained contaminated materials may be an issue. Current airborne monitoring methods can detect any significant agent vapor concentrations. However, new surface analysis technologies may be useful in directing efficient decontamination activities, thus reducing the possibility of airborne agent contamination.

- *Monitoring waste streams.* During operations, agent monitoring can verify that procedures for preventing contamination are effective or assure that decontamination of any agent in or on protective gear and other waste streams are adequate. Waste streams of interest include used protective equipment, processed shell and rocket casings and other packaging materials, and the output of the chemical destruction process itself. Agent monitoring goals include assuring that workers handling waste streams are not subject to acute risks from contaminated material that has uncharacteristically and unintentionally entered the waste stream; assuring that any low-level contamination of the stream does not pose a hazard to workers from long-term, low-level exposure; assuring that any fugitive emissions during transport and disposal of wastes are not of concern; and evaluating the long-term safety of the ultimate disposal and storage methods. It may be important to characterize the mass of agent being exported with a waste stream, even at very low concentrations.

- *Detecting and characterizing incidents of accidental contamination or failure of containment.* The aim is to detect agent contamination before it can spread and be redistributed or cross-contaminate a wider area or commingled material. For such incidents, measuring the degree of contamination and delimiting its spread and extent will be key to efficient handling of contaminated material. Proactive investigation of the amount of spread and redistribution of local contamination from even well-performing routine operations may be wise. Materials that may absorb or trap ambient air agent vapors (such as electrical insulation) may be worth assessing.

- *Providing the basis for segregating contaminated from noncontaminated materials into segregated waste streams.* This issue occurs during process operations but may be more pressing during the facility deconstruction. As noted above, challenges include the possible presence of hidden or occluded contamination that might escape easy detection using vapor monitoring but may be exposed during demolition, collection, and transport of the materials in question. Issues may include the need to screen large amounts of material quickly enough to enable efficient processing into segregated streams on an ongoing, real-time basis and the challenge of efficiently scanning for localized hot spots, where lack of contamination of specific samples may not be sufficient to rule out contamination of the aggregate. Fast and effective detection of agent-contaminated surfaces may provide additional protection to demolition workers against acute releases of hidden reservoirs of material and identify structures that need special care in their demolition.

There are several potential advantages of the new analytical approaches. First, the ability to identify the source of contamination in a complex matrix of potential sources can lead to more efficient detection of residual contamination by agent trapped in occluded spaces. Second, the ability to simultaneously evaluate multiple potential contaminants may also be an advantage of the new measurement capabilities. Third, the ability to detect a source by following an airborne agent concentration gradient with a real-time measurement can lead to more rapid and efficent identification of an agent vapor source and decrease the extent of residual contamination. Fourth, waste streams such as spent activated carbon that are not amenable to analysis of headspace vapors can be interrogated in real time using these new methods.

However, in order to realize the potential benefits listed above, it must first be determined that these new methods have sufficient precision and accuracy to support the required, real-time, low-level-of-agent detection and quantitation in routine practice. What is the sensitivity and specificity of these new methods relative to existing procedures? To what extent can significant residual contamination be identified by these new methods that may not be detectable using existing methods such as headspace vapor analyses? How often can the rapid, in situ capabilities offered by the new methods be used to significantly reduce the time and effort required for necessary routine maintenance, upset response, agent changeover, or closure deconstruction procedures?

## REVIEW OF EXISTING AGENT MEASUREMENT APPROACHES

In reviewing the Assembled Chemical Weapons Alternatives (ACWA) documents, there is little detailed description of the statistical methodologies guiding current agent monitoring and measurement methods. Most available material is presented in the ACWA document *Chemical Agent Laboratory and Monitoring Quality Assurance Plan* (LMQAP) (U.S. Army, 2011b). On page 7 of the plan, the authors state the objective:

Ensure that each operated and maintained monitoring system will indicate by alarm at least 95 percent (%) of the time in the presence of agent at or above the applicable airborne exposure limits (AEL) presented in Section 3.1. A confidence interval of 95% shall be established through the collection of quality data to characterize day-to-day reliability (precision) and validity (accuracy).

However, it is unclear to which statistic the confidence interval applies. In terms of measurement, calibration can be based on as few as three concentration points and is evaluated using a correlation coefficient that will not detect systematic bias. Certification of the method is based on the U.S. EPA method detection limit as described in USEPA 40 CFR Part 136, Appendix B, and involves computing the standard deviation of seven replicate samples spiked at the practical quantitation limit (PQL) and then multiplied by the constant 3.14, which is the 99th percentile of Student's t-distribution with six degrees of freedom. The PQL is typically set at 5-10 times the method detection limit (MDL) and is often called the expected quantitation limit (EQL). This somewhat circular logic (i.e., spike at 5-10 times the analyst's anticipated MDL and then compute the MDL) can yield a wide range of possible detection limits based solely on what a particular analyst expects the result (i.e., the MDL) to be. As will be shown graphically in a later section, the standard deviation is a function of spiking concentration, so the estimated MDL will be a function of the EQL, or, more simply, what the analyst expects it to be. Calibration-based approaches to estimating detection and quantitation limits remove the arbitrariness of this process and provide more statistically rigorous estimates of these unknown quantities. The following section and Appendix D provide a detailed review of calibration and related performance measures that represent a major improvement over common practice.

Chapter 11 of the ACWA LMQAP, "Statistical Analyses," presents statistical approaches to be applied to quality control, calibration, and corrective action data. The discussion is limited and would benefit greatly from further statistical consultation. For example, when computing the statistical response rate at the alarm level, equation 11-10 in the LMQAP assumes that the true population mean and variance are known and the analyte concentration has a normal distribution. However, all that is available from the measurement process is an estimate of the population mean and variance (i.e., the sample mean and variance) from a finite and potentially small number of measurements, drawn from an unknown distribution that is likely to be right skewed and nonnormal. Ignoring the uncertainty in the sample mean and variance will produce incorrect probability estimates. Similar problems for the statistical response rate at the reportable limit (equation 11-12 in the LMQAP) are also present. A more complete discussion of comparing measurements to regulatory standards or alarm limits (e.g., airborne exposure limit (AEL)) is presented in a later section of this chapter and in Appendix E.

## ANALYTICAL MEASUREMENT ISSUES

In this section, the committee considers analytical issues related to how surface (or other) measurement techniques might be qualified as effective for use, including issues of calibration, determination of false-positive and false-negative rates at specified response levels, and the dynamic range of an instrument. The perspective is that the measurement techniques considered in Chapter 4 are designed to offer real-time, spatially resolved quantification of agent contamination. Furthermore, the main value of techniques such as desorption electrospray ionization (DESI), direct analysis in real time (DART), and related techniques may be follow-up evaluation of (relatively larger) areas that have been identified as contaminated through headspace monitoring methods or generator knowledge. For example, the distribution of contamination associated with visually identified quantities of spilled agents can be quickly evaluated.

## Calibration Designs

The most rigorous approach to modeling measurement uncertainty and related detection and quantification limits involves analysis of the calibration function. In order to estimate the parameters of the calibration function and its related properties, a series of reference samples with known concentrations spanning the range of the hypothesized detection and quantification limits are analyzed. Variability is determined by examining the deviations of the actual response signals (or measured concentrations) from the fitted regression line of response signal on known concentration. In these designs, it is generally assumed that the deviations from the fitted regression line are Gaussian distributed; however this is not a requirement for corresponding statistical inference. Both ordinary least squares (OLS) and weighted least squares (WLS) approaches to estimating the calibration function and related uncertainty intervals are commonly used. The advantage of the WLS approach is that it can accommodate nonconstant variance across the range of the calibration function. In general, the magnitude of the variance increases with increasing concentration.

A useful model for the relationship between concentration and variability was originally proposed by Rocke and Lorenzato (1995). This R&L model postulates that there is a small region of constant variability near zero, which then increases linearly with increasing concentration. With a model for the calibration function (e.g., a linear model) and a model for the variance function (e.g., the R&L model), a prediction interval for the calibration function that provides 99 percent confidence for a future measured concentration within the calibration range can be derived. Such intervals provide an estimate of uncertainty in the measured concentration and can also be used to derive a confidence interval for true concentration given the measured concentration. Furthermore, using this statistical methodology, estimates of the detection and quantitation limits can be obtained. The detection limit describes the point at which the analytical method allows a binary decision on whether the analyte is present in the sample, and the quantitation limit describes the concentration at which the signal-to-noise ratio is 10:1 (i.e., a relative standard deviation of 10 percent). All of these statistical estimates document the capabilities of the analytical method and are essential in rigorous environmental monitoring programs. Details of the statistical methodology are presented in Appendixes D and E and are illustrated in the example below.

## CWA Data Analysis Example

To demonstrate the benefit of the statistical data analysis using the calibration methodology described above, the committee analyzed data obtained by H. Dupont Durst and coworkers. These data were summarized in Nilles et al. (2009).[1]

The spectral area data (for agent and an isotopically labeled agent internal standard) were acquired with a time-of-flight mass spectrometer. Measurement samples were obtained by dipping a closed glass capillary tube into agent solutions (GA, GB, VX, and HD) and then inserting it into the DART ion source. Data at multiple concentrations in low (20 ppb), medium (200 ppb) and high (1 ppm) ranges were available for VX in deionized water and VX in 2-propanol, and GA, GB, and HD in methylene chloride. Each calibration generally included seven replicates at each of six different concentrations. For purpose of illustration, the committee selected the middle concentration range because the instrument responses are well differentiated across the concentrations in all matrices. VX in deionized water and VX in 2-propanol were also reasonably well differentiated at the lower concentration calibration, but GA, GB, and HD in methylene chloride were not. Quite similar results were obtained for VX in both low- and medium-concentration calibrations. The R&L model was used to characterize the variance function. This illustration focuses on VX in deionized water. All computations were performed using the Detect software (Discerning Systems, Vancouver, Canada). Figure 5-1 displays the estimated calibration function for VX in deionized water. Figure 5-2 displays the estimated relationship between variability and concentration. Figure 5-3 displays the relationship between the percent relative standard deviation and concentration.

This analysis reveals that the binary detection decision for VX in deionized water can be made at 2.7 ppb with 99 percent confidence, with a detection limit of 5.4 ppb. Quantification (i.e., the concentration at which the signal-to-noise ratio is 10:1) is possible at concentrations at or above 11.2 ppb. Applying the same methodology to the lower calibration range yielded critical level $L_C = 2.5$ ppb, detection limit level $L_D = 4.9$ ppb, and quantitation level $L_Q = 10.3$ ppb, indicating that the estimated quantities are relatively robust to the selection of the calibration range.

For comparison purposes, the committee also examined the data for VX in 2-propanol. Here, it obtained $L_C = 6.7$ ppb, $L_D = 13.4$ ppb, and $L_Q = 28.0$ ppb. These results suggest a substantial matrix effect, indicating that it is more difficult to detect and quantify VX in 2-propanol than in deionized water.

For GA, GB, and HD in methylene chloride, the committee obtained the following:

GA:  $L_C = 16.4$ ppb, $L_D = 32.8$ ppb, and $L_Q = 67.0$ ppb.
GB:  $L_C = 22.6$ ppb, $L_D = 45.0$ ppb, and $L_Q = 92.8$ ppb.
HD:  $L_C = 11.0$ ppb, $L_D = 22.1$ ppb, and $L_Q = 46.7$ ppb.

---

[1] The committee used an extended dataset provided by H. Dupont Durst, Edgewood Chemical Biological Center, CMA.

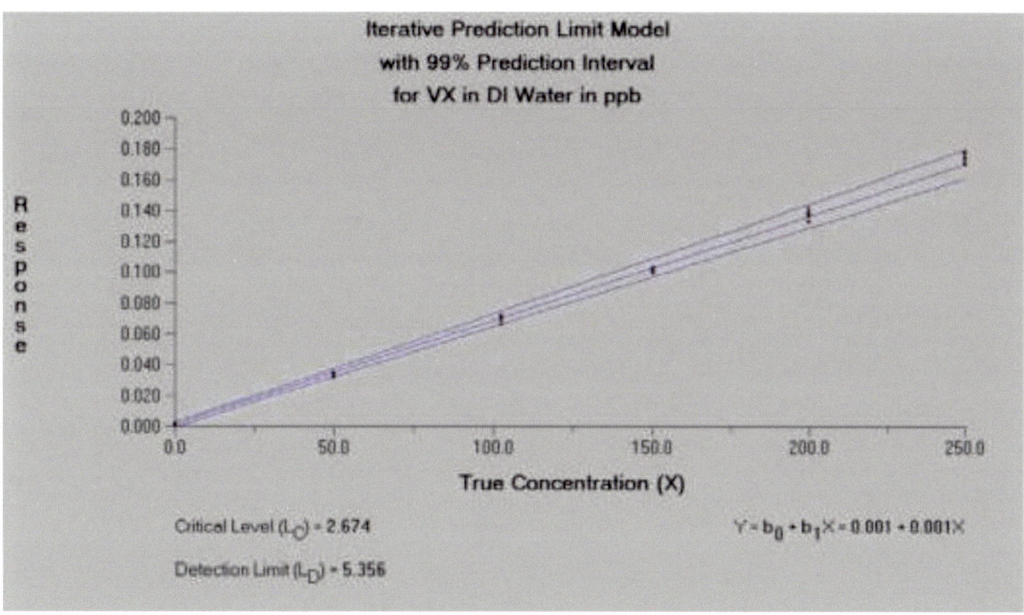

FIGURE 5-1 Estimated calibration function for VX in DI (deionized) water. SOURCE: Constructed from data provided by H. Dupont Durst, Edgewood Chemical Biological Center, CMA.

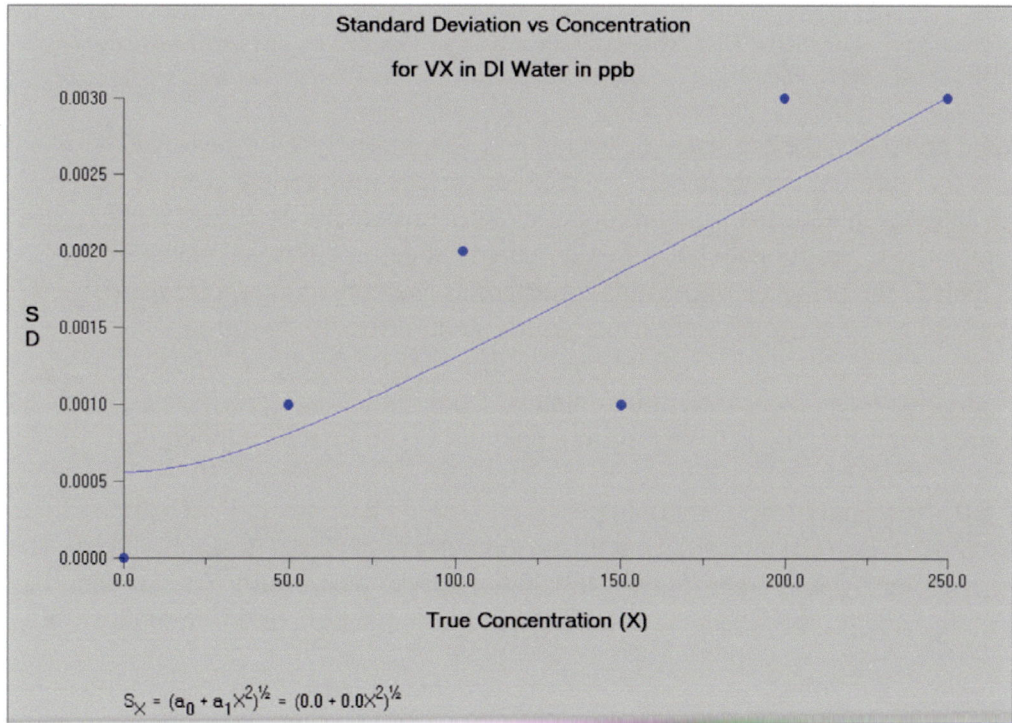

FIGURE 5-2 Estimated relationship between variability and concentration for VX in DI (deionized) water. SOURCE: Constructed from data provided by H. Dupont Durst, Edgewood Chemical Biological Center, CMA.

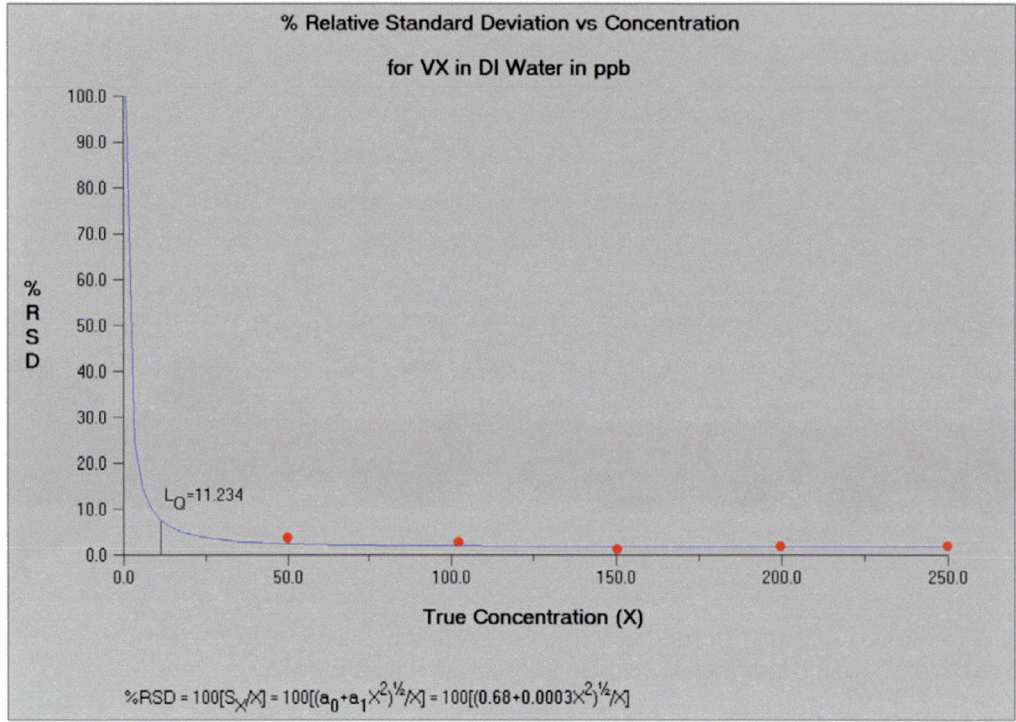

FIGURE 5-3 Relationship between the percent relative standard deviation (%RSD) and concentration for VX in DI (deionized) water. SOURCE: Constructed from data provided by H. Dupont Durst, Edgewood Chemical Biological Center, CMA.

It should be noted that the estimated detection and quantitation limits described here may not accurately represent expected performance for surface monitoring. Results obtained for single analytes in aqueous or liquid organic matrices may not be achievable in real-world applications. Additional challenges for chemical demilitarization surface data analyses include the presence of potential interferents; heterogeneous solid matrix properties; and heterogeneous analyte distributions and difficulty in introducing any internal standards. In the absence of internal calibration standards, the mean concentration for each physical sample (obtained from replicate samples) can be used as the reference and the methods described in Appendix D can then be applied.

**Finding 5-1.** New analytical methodologies (e.g., DART) have been demonstrated for relevant nerve agents (GB and VX) and blister agents (HD, HT, H) in simple liquid matrices (deionized water, isopropyl alcohol, and methylene chloride). Similar data for relevant surfaces (e.g., metal, concrete, activated carbon, plastics, and iron oxide) are sparse.

**Recommendation 5-1.** The use DART, DESI, or related new analytical methodologies for surface area measurements at the Pueblo Chemical Agent Destruction Pilot Plant or the Blue Grass Chemical Agent Destruction Pilot Plant requires that the quality of measurements be determined and related calibration studies be performed for relevant matrices.

**Finding 5-2.** Good precision and accuracy for DART techniques have been established for liquid matrices through the use of an internal standard. Challenges for development of internal standards for surface measurements include the presence of potential interferents; heterogeneous solid matrix properties; and heterogeneous analyte distributions and difficulty in introducing any internal standards. Development of internal standards for more homogeneous liquid and gas-phase matrices is more straightforward. The development of an internal standard may or may not be practical for some surface analysis applications at the Pueblo Chemical Agent Destruction Pilot Plant or the Blue Grass Chemical Agent Destruction Pilot Plant. In the absence of an internal standard, the precision of the quantitative measurements may decrease.

**Recommendation 5-2.** In the absence of an internal standard for surface measurements, the uncertainty in the measurement technologies (e.g., DART and DESI) should be established.

## COMPLIANCE MONITORING

An important aspect of any monitoring program is demonstrating compliance with closure or remediation objectives. In the current context, the null hypothesis is that all surfaces, machinery, materials are contaminated and that the null hypothesis of contamination is rejected only when there is overwhelming evidence that the material in question is not contaminated. This specification of the null and alternative hypotheses leads to consideration on an upper bound on the true concentration of the analytes of interest, for example an upper confidence limit (UCL) for the mean concentration. To the extent that the UCL is beneath a health-based standard, cleanup objective, or worker safety threshold, it can be concluded that the material has been decontaminated, or that contamination was not present to begin with. This line of thinking can potentially lead to significant advantages in closure and related monitoring. If materials can be screened using new technologies, then headspace-type sampling and related analytical methods may not be required.

As an example, consider a surface such as a wall, which can be divided into grids. Wipe samples of each grid can then be used to produce a spatial distribution of analyte concentrations over the entire surface. If the spatial distribution is reasonably homogeneous (e.g., a nonsignificant random grid effect), then the average concentration of the grid measurements and the corresponding UCL can be estimated (based on an appropriate distribution for the chemical concentrations), and the UCL can be used to make a determination if further analytical or remedial work is required. Appendix E provides detailed parametric (normal, lognormal, gamma) and nonparametric methods for computing the appropriate UCL for relevant applications. If the spatial distribution is not homogeneous, then more local characterization of the concentration distribution can be performed and potential hot spots located and remediated until the distribution becomes homogeneous and an areawide determination can be made.

**Finding 5-3.** The published protocols for statistical procedures for compliance monitoring by the Assembled Chemical Weapons Alternatives program (i.e., in the draft *Laboratory Monitoring Quality Assurance Plan*, Rev. 0, April 4, 2011) contain insufficient detail to provide guidance for compliance monitoring. The ambiguity of current publicly available documentation suggests that quantification of statistical variability has the potential to be inaccurate. Such inaccuracies may result in difficulties such as unnecessary destruction of uncontaminated waste and/or failure to identify contaminated waste.

**Recommendation 5-3.** The Assembled Chemical Weapons Alternatives program should reexamine existing protocols planned for compliance monitoring at PCAPP and BGCAPP and means for incorporating increased statistical rigor in the assessments to be performed.

## STATISTICAL SAMPLING ISSUES

### Measurement Bias, Precision, and Detection Limits

The "Analytical Measurements Issues" section in this chapter describes statistical approaches to understanding the information about a measurand (the quantity of interest) that can be inferred from one or more measurements. These discussions focus on measurement error as a source of random variability in measurement. With respect to the uncertainty associated with measurement error, calibration primarily refers to investigation of the relationship between a hypothetical "noiseless measurement" and the quantity being measured. In practice, calibration usually also refers to "correction" of a measurement system to eliminate measurement bias. In contrast, measurement precision refers to unreproducible random measurement error that leads to variation among multiple measurements of the same quantity.

In this section, it is assumed that the measurement systems of interest have been calibrated, so that bias has been eliminated. Conceptually then, a measurement can be thought of as the sum of the measurand and a random measurement error that has an expected (or long-run average) value of zero. The standard deviation of this random noise characterizes the precision of the measurement instrument (where a larger standard deviation corresponds to poorer precision). The precision associated with a measurement device can be improved by using the device to obtain multiple independent determinations of the same measurand and reporting the average of these as the measurement. In contrast, the bias associated with a measurement device cannot be reduced by repeated measurement; this fact underscores the critical importance of adequate calibration of any measurement system (Vardeman and Jobe, 1999).

Most measurement instruments used to determine the quantity or concentration of a substance also have a detection limit, a measurand value below which the instrument cannot discriminate true zero measurands from small positive values. When a measurement will be used alone (i.e., not along with other measurements in a more comprehensive analysis) and the detection limit is less than any applicable action level, the detection limit has little or no practical implication. However, when the detection limit is not small relative to action levels, and/or multiple measurements are to be analyzed together, appropriate statistical treatment of the data can become more complicated. For example, simply substituting zero for the detection limit value for below-detection-limit readings can easily lead to seriously biased estimates and flawed conclusions. In these cases, appropriate censored data methods should be used to assure valid results (Helsel, 2004).[2] However, while the more statistically rigorous analyses described above are more complicated than simple imputation methods, they may be captured in straightforward software routines to analyze ambient ionization data in real time and present evaluated data to the operator to guide further measurements.

**Finding 5-4.** Reliable real-time computer programs able to interpret real-time chemical analyses enable instruments (generally using proprietary software) to convert intensity measurements to concentrations of agents and potential interferents.

**Recommendation 5-4.** Instrument software for use with ambient ionization mass spectrometry should be reviewed to ensure that it meets appropriate validation and verification criteria. This software should be tested by using simulated data to test different measurement scenarios (e.g., all data below detection limits, at detection limits, mixtures, hot spots, and so on).

## Measurement Basis

The measurement technologies used by ACWA, and those additional technologies that might also be considered, can be applied with different sampling strategies that correspond to substantially different definitions of "measurand." In particular,

- Miniature continuous air monitoring system (MINICAMS) and depot area air monitoring system (DAAMS) measurements represent agent concentrations in air, integrated over fixed time intevals.
- DART and DESI techniques can be used to provide measurements of spatially resolved concentrations on surfaces.

Air concentration values have the most obvious connection to worker risk and so provide the basis for most existing standards and action values. However, where the challenge is to identify the location of quantities of adsorbed, absorbed, or trapped condensed-phase agent for decontamination, air concentration values are indirect and spatially indistinct indicators of the quantities of greatest interest. Conversely, while surface measurement techniques may accurately characterize the degree of contamination at a spatial point, such measurements are not efficient as a basis for screening a larger area or volume for contamination due to the number of such measurements that would be

---

[2]In statistical modeling, measurement data that may take on numerical values but may also result only in an indication that a threshold value has been exceeded (e.g., "below detection limits" or "above saturation level") are termed "censored." Statistical methods specifically designed to deal with such mixed data values are called "censored data methods."

required. As discussed in the section "Hot Spot Detection" later in this chapter, the joint use of measurements representing different sampling modes may be an effective strategy in some settings.

In addition, surface measurements representing a larger basis than a near-point basis may also be achieved in some cases by some form of composite sampling—the collection and physical combination of material from several locations or over a continuous physical area—followed by one or more measurements of the combined sample. In particular, "wipe sampling" refers to a process in which a fabric matrix is passed across an extended area of the surface to be monitored and is then processed before a measurement is made. Such techniques can improve the efficiency of sampling activity, especially when the measurand is substantially below the action limit over most of the area to be characterized, because it may be possible to "clear" extensive areas using relatively few individual measurements.

However, when spatially integrated (e.g., composite) sampling is used, some thought must be given to how the resulting measurements should be related to the action levels. For example, if the intent of the standard is that *no* individual point on a surface have agent at a concentration above a specified level, wipe sampling may be only indirectly applicable if the concentration in the collected sample represents the average over the area wiped.[3] In addition, there can be losses in efficiency associated with the process used to collect material; see, e.g., Verkouteren et al. (2008).

**Finding 5-5.** In some cases, analysis of direct surface and/or materials wipe sampling may complement or replace vapor screening level analysis, allowing more efficient and cost-effective closure operations.

**Recommendation 5-5.** If direct surface and/or materials wipe sampling analysis methods are adopted, appropriate statistical methods for characterizing the extent of contamination of surfaces, machinery, and/or materials should be employed.

## Spatial Modeling

In simple analytical applications where the measurement device is known to be well calibrated, it may be sufficient to limit analysis of measurement variation to random measurement error, considering only one measurand at a time. However, ACWA applications may also involve the use of one or more measurement methods within a spatial area[4] to be monitored or characterized. For example, Scenario 3B (see Chapter 3) refers to demilitarization protective ensemble (DPE) suit entries into contaminated areas with the goal of identifying (the spatial location of) a source.

---

[3] In this sense, "average" refers to the mechanical process of collecting material over multiple locations or a contiguous surface. Single or repeated measurements of such combined samples may be expected to yield values that are between the extreme (high and low) concentrations that exist over the area sampled.

[4] A two- or three-dimensional region, such as a floor or room, over which agent concentration (on the surface or in the air) may vary. It is important to understand this variation, especially the more specific locations (in the spatial area) where concentration is unacceptably high.

In this context, a portable ambient ionization instrument, as described in Chapter 4, that produces a measurement of surface concentration of an agent specific to a location (actually, a small volume or area of surface, but here treated ideally as a point) within the region of interest may be used to collect data that can support the calculation of a contour map of concentration throughout that region. (For this discussion, the possibility that the measurand might change with time is ignored.) Regardless of the measurement technology used, it is clearly impossible to acquire a measurement at every spatial point of interest; at best, measurements taken at some collection of locations, or reflecting a "mechanically averaged" value over an area using a wipe, can be collected. As a result, there will be uncertainty stemming from spatial variability—the actual variation of the measurand across the area being studied—in addition to the uncertainty associated with measurement imprecision at any point. Specification of the nature of spatial variability is a critical step in deciding how samples should be collected, and how the resulting data should be analyzed for the purpose of monitoring or characterizing agent concentration in an area.

One widely used approach to spatial modeling describes the measurand as a stochastic spatial function of location within which concentration values are regarded as draws from a probability distribution (Cressie, 1993). Rather than being specified as independent draws at each location, they are modeled within a framework that allows statistical dependence, with stronger dependence for pairs of spatial locations that are separated by shorter distances.[5] The result is that measurements collected at a fixed collection of spatial locations can be used to make useful predictions of concentration at other locations. One specific modeling method of this type, kriging,[6] is widely used in environmental characterization studies and was employed first in mining applications where measurements at a collection of "core" sites were used to predict potential ore yield throughout an area or volume to be explored (Journel and Huijbregts, 1978).

Scenario 3B presents one possible setting for spatial modeling in the context of ACWA activities. If a spill has occurred in a work area and has been spread spatially by accidental contact or previous unsuccessful cleanup efforts, a spatially coherent (even if irregular) distribution of agent concentration may result. Sampling procedures that can efficiently acquire the measurements necessary to support reliable characterization of such agent deposits during DPE entries may be developed based on stochastic spatial models of agent concentration, so that areas requiring decontamination can be quickly identified.

Because the collection of measurements taken over a spatial domain is likely to be skewed in this context (i.e., many or most values are below detection, and perhaps a few are larger), the fidelity of spatial modeling may be improved by a nonlinear, monotonic transformation (such as a logarithm) of the measurand data. No information is lost in

---

[5]In statistical modeling, two random quantities are "independent" if their values are completely unrelated, that is, if knowing the value of one does not influence what can be inferred about the other. "Dependent" random quantities, while each being subject to unpredictable noise, are related in such a way that knowing the value of one reduces the uncertainty in the other. For example, pairs of random quantities that are correlated (higher values of one are generally associated with either higher or lower values of the other) are dependent.

[6]Kriging is a statistical technique, originally developed for geological applications, that is used to estimate a continuous map of some quantity (e.g., a concentration) from measurements of that quantity taken at a discrete set of locations.

such a transformation, but the degree of spatial variability is typically more consistent in the transformed data, across the modeling domain. When modeling is done on a transformed scale, the resulting model (or "contour map") can be reverse-transformed to the original, physically meaningful scale.

## Sampling Plans for Spatial Modeling

There is no one optimal spatial sampling plan, even when a spatial variation model has been fully specified, because plans that are good for some purposes may be entirely inappropriate for others. For example, measurements may be collected within a defined spatial domain in order to support the following:

- Estimation of the parameters in a spatial model,
- Estimation or prediction of the measurand at a specified selection of locations not included in the sampling plan,
- Estimation or prediction of the integral or average of the measurand over the entire region of interest or a defined subregion,
- Estimation or prediction of the largest measurand within the region of interest, and
- Estimation or prediction of the location within the region of interest where the measurand is largest.

For any specified statistical model, each of these criteria may lead to different "optimal" sampling plans. However, within any specific application and for almost any realistic characterization goal, most reasonable sampling plans include:

- Measurements taken at a sufficient number of locations to provide reliable characterization of the spatial variation of the measurand across the region of interest and
- A sufficient number of replicate measurements at some locations to determine the magnitude of measurement errors being encountered or to validate the assumed precision of the instrument.

While the particular details of an efficient sampling plan must depend on specific goals and the details of the statistical model to be used, most reasonable models will lead to sampling plans with these two characteristics. Two general approaches to spatial sampling are briefly described below.

*Fixed Sampling Plans*

The simplest spatial sampling plans specify a collection of locations at which measurements are to be taken, and the number of replicate measurements to be taken at each location. For example, Scenario 3E anticipates that possible contamination over a concrete surface may need to be characterized at facility closure in order to minimize the extent of scabbling. In this example, the floor of a 10 m × 10 m room might be sampled at 100 points, selected to cover the surface. These 100 points might be chosen randomly; however, randomly selected points often display clustering behavior. That is, some contiguous subregions may contain several selected sampling locations while others of similar size contain few or none. An approach that is often more reasonable is to use a distance-based measure to ensure that the 100 points selected cover the space as evenly as possible. Uniform discrepancy sampling designs (Fang et al., 2000) and maximin distance sampling designs (Johnson et al., 1990) are two approaches that can be more effective than simple random sampling of locations.

Generally, replicated measurements should be collected for at least some of the specified locations. In the committee's hypothetical example, it might call for three replicate measurements to be made at each of the 100 locations in the room. Hence the overall sampling plan calls for 300 measurements in total. Given this restriction to 300 measurements, one might ask whether it would be more effective to collect four replicates at each of 75 locations (more replicate sampling at fewer spatial points) or two replicates at each of 150 locations (less replicate sampling at more spatial points). Determining an optimal balance requires some a priori generator knowledge of the relative sources of uncertainty associated with spatial variation of the measurand and with measurement error and also with the type of statistical model to be used for characterizing variability. In general, for measurement systems that are more noisy (i.e., that imply poorer measurement precision) relative to the actual variation in the measurand throughout the region, plans that sample at fewer distinct points and include more replicate measurements are preferred.

*Sequential Sampling Plans*

In contrast to fixed sampling, sequential sampling plans are designed so that later sampling locations are influenced by early measurements. Because one goal of sampling in this context is to identify subregions in which concentration is high, sequential sampling plans that use early measurements to get a rough idea of where these may be and carry out later measurements in the subregions that appear to be critical, can be more efficient than fixed sampling plans. The instruments described in Chapter 4 are rapid response (approximately 1 sec), making sequential sampling viable. As an alternative to the hypothetical fixed plan using 300 measurements described above, the following might be used.

A first stage of surface sampling might be carried out at 50 uniformly selected sites, with two replicate measurements at each point. Based on these 100 measurements and an acceptable statistical model, a "contour map" of agent concentration predictions of the measurand could be computed for the floor of the 10 m × 10 m room. This analysis will also yield standard errors, or confidence intervals, for the concentration at any point in the room. Some areas may be clearly below the applicable action value (i.e., with an upper confidence limit less than the action level), and some may be clearly above it (with a lower confidence limit greater than the action level). From a practical standpoint, there is little value in further sampling from either of these types of regions. Instead, a second stage of sampling might be collected at 50 additional locations uniformly spread throughout the "ambiguous" regions, where upper and lower confidence limits of concentration are on either side of the action limit, perhaps again with two replicates at each location. A second analysis using all 200 data values collected in stages 1 and 2 will yield a more precise prediction of concentration in the areas that were initially difficult to classify, owing to the increased sampling intensity in these areas during the second phase of sampling. This process could then be extended to a third round of sampling, focused on regions that remain ambiguous based on the updated analysis.

Sequential sampling can be operationally impractical in applications where the analytical processing required for each measurement is substantial, or when the required data analysis between samples is complex. In contrast, sequential sampling appears to have great potential value when used with ambient ionization mass spectrometry because the latter provides location-specific data with very fast turn-around. Further, while interim formal statistical analysis may be needed in some cases, near-real-time results from these techniques will make it possible for workers to collect informal but informative sequences of measurements (without formal interim analysis) based on the characteristics of the sampled environment and the emerging patterns in the data.

**Finding 5-6.** The use of statistical sampling will improve agent contamination detection and quantitation. For near-real-time measurement technologies, sequential sampling may be particularly valuable. Specific sampling plans will depend on the geometry of the contaminated area, contaminant spatial variability, and the goal of the measurement process.

**Recommendation 5-6.** The Assembled Chemical Weapons Alternatives (ACWA) staff should have access to sufficient statistical expertise to develop effective sampling protocols for any application of ambient ionization monitoring. Once the resulting expert sampling protocols have been developed, ACWA headquarters monitoring staff or their contractors should then proceed to develop detailed standard operating procedures to guide monitoring technicians.

## Hot Spot Detection

Another agent deposition pattern that may be especially relevant in the ACWA context could be formulated to describe one or a few hot spots in a region that otherwise has no agent or only a very low concentration of agent. Strictly speaking, this could also be regarded as a spatial pattern, but it is one in which there is relatively little spatial coherence and so relatively little opportunity to interpolate the measurand values at unsampled locations. Instead there are a few sparsely distributed but highly contaminated regions, and throughout the remainder of the area, contamination is comparatively low. For practical purposes, the important quantities to be determined are the elevated level and location of contamination.

In the context of Scenario 3D, it may not be sufficient to describe the occluded spaces using simple spatial coordinates. Cracks in concrete, pump chambers, and quantities of absorptive material may be pinpointed in space, but the specific characteristics of these potential agent sinks and their patterns of operation may produce more helpful information than their physical coordinates. Here, potential hot spots coincide with these physically identifiable entities, and one or a few point-wise measurements within each may be a sufficient characterization.

The distinguishing characteristic of this model is the disconnected way in which it characterizes the contamination at each location, even within a spatial domain; essentially nothing is being assumed about spatial structure.[7] That is, the model provides little or no basis for making any claim about what might exist at any unsampled locations. More specifically, there is no statistical basis for predicting or estimating the contamination at an unsampled location, no matter how close it is to the locations actually sampled. With sufficient sampling, including replicate measurements at some points, the model parameters (such as the mean and standard deviations representing measurement errors and the spatial variability of the contamination) may be estimable, and through this it may be possible to predict, for example, the proportion of the region of interest that exhibits significant contamination. But there may be little or no basis for identifying the sub-regions that may be most problematic.

From a modeling standpoint, hot spot phenomena are much more difficult to characterize than more gradually varying spatial patterns because of the noted lack of spatial coherency. Specifically, when hot spots comprise a relatively small area of volume compared to the region that must be screened, their location cannot be inferred by spatial interpolation from locations where the measurand is low. For practical purposes, a hot spot can be detected only by a measurement reflecting the concentration at the unknown location of the hot spot. For this purpose, sequential sampling plans that utilize multiple sampling bases may be most effective, as described below.

## Sampling Plans for Hot Spot Detection

The fixed and sequential sampling plans described above can sometimes be useful in detecting isolated hot spots of agent concentration in spatial domains. However, if the hot spots are small in volume or area and the background concentrations do not "ramp up" to these elevated levels at nearby locations, the likelihood of identifying a hot spot with a spatially distributed collection of near-point measurements will often not be great.

Locating isolated agent deposits may be more effectively accomplished by a sequential strategy in which early samples are made on a broad area or air volume basis (e.g., headspace determinations), followed by measurements of wipe samples taken from limited surface areas, and finally with spatially resolved measurements.

The efficiency of sampling to identify hot spots may be improved substantially through the use of reliable generator knowledge in the temporal ordering of measurements. Ordered sampling suggests that there is some prior reason for making measurements at some locations earlier than at others. For example, in occluded space

---

[7] The kinds of spatial structure referred to might include continuity of the quantity of interest as a function of location, or any other characteristic that suggests a systematic connection between location and that quantity. In contrast, here the committee is discussing a model for which all pairs of spatial locations have the same relationship; for example, distance has no relationship to how much difference might be expected in the quantity of interest.

surveys (Scenario 3C), operation patterns and historical records will often suggest that some locations are more likely sources of agent than others. Arranging the order of measurements so that the elevated agent concentration is more likely to be found earlier rather than later can shorten inspection, at least in cases where a single reservoir can be tentatively assumed to be the source of contamination.

**Finding 5-7.** The successful application of any measurement technology is a function not only of its capabilities, but also of the ultimate use of the data generated. The context and purpose of the measurements will determine the sampling scheme, precision, and accuracy required.

A synopsis of the major issues identified in Chapters 1 through 5 is presented in the next and final chapter.

# 6
# Report Summation and Recommendations

As outlined in Chapter 1 of this report, the U.S. Army's Chemical Materials Agency and its predecessor organizations have been engaged in the demilitarization of the nation's stockpiles of chemical weapons for over a quarter of a century. CMA recently completed destruction of the chemical agents and associated munitions stored at six of eight continental U.S. storage facilities as well as chemical weapons deployed overseas, which were transported to Johnston Atoll, southwest of Hawaii, and demilitarized there. These CMA activities have successfully destroyed 90 percent of the nation's chemical weapons inventory

The remaining 10 percent of the chemical weapons stockpile is stored at two remaining continental U.S. depots, in Lexington, Kentucky, and Pueblo, Colorado. Their destruction has been assigned to a separate U.S. Army organization, the Assembled Chemical Weapons Alternatives (ACWA) Element. The last two chemical weapons disposal facilities, the Blue Grass and Pueblo Chemical Agent Destruction Pilot Plants, are currently under construction. ACWA is charged with destroying the Blue Grass and Pueblo stockpiles without using the multiple incinerators and furnaces used at the five CMA demilitarization plants that dealt with assembled chemical weapons—munitions containing both chemical agents and explosive/propulsive components.

The CMA disposal facilities that processed assembled chemical weapons used an array of furnaces both to destroy drained chemical agents and to decontaminate or destroy other agent-contaminated munitions components and secondary waste materials. However, the two ACWA demilitarization facilities are congressionally mandated to employ noncombustion-based chemical neutralization processes to destroy chemical agents and will not have large furnaces to decontaminate or destroy secondary waste materials. This constraint motivates an interest in analytical methods that can quickly and reliably identify and characterize agent-contaminated materials. Detecting and characterizing agent-contaminated structural surfaces are also a priority, both during agent changeover operations (BGCAPP only) and during facility closure activities, when agent disposal facilities must be decontaminated before demolition (both BGCAPP and PCAPP). Since chemical weapons disposal operations are currently not expected to start at PCAPP until 2015 and at BGCAPP until 2020, there is at least a 2- to 7-year window to assess, develop, and procure advanced analytical technology that could be used in both agent processing and closure activities at each site.

Currently available methods to monitor chemical agent contamination of both secondary waste and structural components (initially discussed in Chapter 2 and more thoroughly in Chapter 3) were developed at CMA disposal facilities and will be adopted for use at the ACWA facilities. While they have allowed safe waste processing and closure activities, these methods are time consuming and indirect, generally relying on vapor-phase agent measurements over confined surfaces, rather than direct detection of surface contamination.

Chapter 3 also summarizes the characteristics and estimated magnitudes of potentially contaminated secondary waste streams that will be generated during both process operations and closure activities at BGCAPP and PCAPP. Based on current understanding of probable agents/munitions processing and closure activities at the ACWA facilities, six scenarios developed by the committee are presented in which real-time agent contamination measurements on surfaces or bulk materials might allow more efficient, and possibly safer, operations.

The recent rapid development of ambient ionization mass spectrometric techniques for real-time surface and bulk materials analyses is reviewed in Chapter 4. An assessment of the capability of these techniques to provide highly sensitive and specific real-time measurements of the chemical agents relevant to ACWA demilitarization activities is presented, and implementations of the technologies relevant to selected Chapter 3 scenarios are evaluated, in Chapter 4. The ability of these methods to perform real-time chemical agent vapor concentration measurements is also explored.

Chapter 5 assesses the statistical measurement challenges inherent in both current vapor-phase chemical agent monitoring and potential ambient ionization surface/bulk agent contamination measurements. Statistical constraints on real-time sampling methods pertinent to the agent contamination scenarios presented in Chapter 3 are discussed and assessed in Chapter 5. Two appendixes associated with Chapter 5 present statistical methods that can be used to develop and evaluate potential chemical agent contamination measurement strategies.

The remainder of this chapter summarizes salient points from Chapters 2 to 5 in conjunction with a reiteration of the findings and recommendations presented in those earlier chapters, and it concludes with an assessment of the potential value of ambient ionization mass spectrometry for prospective ACWA chemical weapons demilitarization activities.

## FINDINGS AND RECOMMENDATIONS

The initial finding of this report (2-1) points out the indirect nature of the methods that CMA developed for detecting and assessing chemical agent contamination of materials, and that ACWA plans to use during agent processing and plant closure operations. Closely related Finding 2-2 recognizes a recent CDC recommendation that CMA establish a health-based surface agent contamination hazard level for use with wipe samples of agent-contaminated surfaces interrogated during closure activities. The committee observes that such a standard, coupled with new capabilities for real-time direct surface contamination measurements, could also serve ACWA well.

**Finding 2-1.** The prevalent Army demilitarization activity methods of detecting materials' surface contamination involve enclosing materials and monitoring headspace agent concentrations. These are indirect methods that can determine if significant levels of agent are present in the enclosed volume; surfaces are not directly monitored. However, vapor detection does not identify the location nor quantify the level of contamination on surfaces within the test volume.

**Finding 2-2.** No CMA or ACWA standards have been established for surface contamination similar to the airborne agent concentration exposure limits, from which vapor screening levels have been adopted. If accepted by the CDC and relevant state regulators, a health-based agent-contaminated surface hazard level measured in mass per unit area by a new, direct surface contamination measurement technology and suitable agent-contaminated surface calibration standards could be useful in clearing secondary waste materials during ACWA disposal operations and/or structural materials during closure. However, reliable agent-contaminated surface calibration standards may be difficult to produce.

Based on its review in Chapter 3 of available estimates of anticipated secondary waste production at both PCAPP and BGCAPP, the committee notes that the two sites currently tabulate anticipated waste streams using disparate waste category designations.

**Finding 3-1.** The waste category designations used for tabulating waste streams at the Pueblo Chemical Agent Destruction Pilot Plant and the Blue Grass Chemical Agent Destruction Pilot Plant differ, thus making waste management comparisons between the two facilities difficult. For example, at one site the waste quantity estimates list waste demilitarization protective ensemble suits separately, but at the other such waste is included in halogenated plastic waste.

Accordingly, the committee recommends:

**Recommendation 3-1.** The Program Executive Officer for Assembled Chemical Weapons Alternatives should consider implementing a uniform set of waste category designations for use at both the Blue Grass Chemical Agent Destruction Pilot Plant and the Pueblo Chemical Agent Destruction Pilot Plant to facilitate the transfer of knowledge and lessons learned between sites.

After also reviewing the major role that demilitarization protective ensemble (DPE)-suited entries into agent-contaminated weapons processing areas are expected to play in pacing ACWA facility weapons disposal, the committee finds:

**Finding 3-2.** Any new monitoring method that could efficiently and reliably locate and quantify agent contamination may make decontamination activities more efficient by:

- Enabling faster identification of leaking munitions and decontamination of machinery, potentially reducing the number and/or duration of DPE-suited entries during normal plant operations, agent changeover periods, and closure activities;
- Reducing the total amount of secondary waste;

- Speeding waste disposal; and
- Minimizing worker exposure.

Consideration in Chapter 3 of the need to identify and assess potential agent contamination in occluded spaces during agent changeover or closure activities led the committee to find:

**Finding 3-3.** A local, real-time agent monitoring system capable of monitoring surfaces might enhance the effectiveness of occluded space survey teams by identifying problematic occluded spaces and identifying other sources of contamination, possibly reducing the time necessary to conduct agent changeovers or facility closure.

After reviewing (in Chapter 3) the likelihood of agent contamination of porous materials such as concrete, as well as prior NRC committee recommendations (NRC, 2007, 2008a), the committee also finds:

**Finding 3-4.** Materials with inherent porosity can readily adsorb or absorb agent and present a monitoring challenge for headspace vapor measurement methods.

Chapter 4 presents a thorough overview of the measurement principles, technology variations, and analytical capabilities of recently developed ambient ionization mass spectrometry techniques. Based on information about and analyses of the technology's capabilities, the committee finds:

**Finding 4-1.** When compared to existing vapor monitoring (DAAMS and MINICAMS) measurement strategies, ambient ionization mass spectrometry provides the following capabilities for detection and quantitation of chemical agents and their degradation products (see also Table 4-3):

- Exceptional sensitivity and selectivity.
- Direct measurements from vapor, liquid, or solid samples.
- Real-time measurement with minimum response times of milliseconds to seconds.
- Multiagent detection capability and degradation product monitoring.
- Measurements of variations in spatial and temporal concentration.

**Finding 4-2.** DART, DESI, and other emergent ambient ion sources can be considered for surface analysis. A range of different technologies can be employed for liquid and ambient vapor ionization. Both DESI and DART are able to sample and identify molecular species on surfaces in real time (less than 5 sec per measurement).

**Finding 4-3.** DESI requires a solvent, usually an organic or organic/water mixture, with a few percent acid to enhance protonation of the target molecules. A high-velocity gas flow concentric with the electrospray tip directs charged droplets toward the sample. This plume can easily scatter liquid and solid material. Further dispersal of chemical agent that might lead to contamination of adjacent surfaces, especially the mass spectrometer, is not a desirable result.

**Finding 4-4.** DART, using a controlled flow of heated gas typically comprising helium or a nitrogen/helium mixture, provides the capability to heat, evaporate, and subsequently ionize the target species with modest target agent dispersal. Furthermore, it is possible to vary the temperature of the flowing gas to carry out temperature programmed desorption, observing in turn species with a wide range of volatilities. In addition to surface sampling capabilities, the DART source is an efficient atmospheric pressure chemical ionization source and hence can ionize and detect species at low concentrations in the vapor phase.

**Finding 4-5.** The platform configuration most likely to satisfy the analytical needs put forward in the various scenarios in Chapter 3 for different waste streams consists of a cart-mounted or handheld mass spectrometer equipped with a modified interface to accommodate a special remote sampling wand, a surface ambient ionization source combined with a vapor ambient ionization source, and any sampling accessories. Ambient ionization mass spectrometry systems backed by an uninterrupted power supply will allow portability between different rooms or site areas without breaking vacuum. Careful attention to instrument shielding and sampling wand design and implementation can reduce the possibility of agent contamination during instrument use.

**Finding 4-6.** At the highest levels of sensitivity, detection of chemical agents requires high mass resolution to distinguish the agents from isobaric trace species with the same nominal mass. A minimum resolution of 10,000 (m/$\Delta$m, where $\Delta$m is the full width at half maximum) is required for this purpose. This value can be accomplished with an orthogonal sampling time-of-flight (TOF) mass spectrometer incorporating a reflectron. High-end instruments of this type can provide mass resolution 1.5-5 times this value and would be preferable if not for their cost and size. Selective reaction monitoring, in which an ion structure is confirmed by specific fragmentation pathways, can avoid false positives for chemical agent identification without requiring high mass resolution. While this can be accomplished with newer TOF-TOF instruments, the suggested mass analyzers are either a triple quadrupole or a linear ion trap.

These findings inform the following recommendations:

**Recommendation 4-1.** While both DESI and DART have been demonstrated to have excellent sensitivity for detecting chemical agents in liquids and on a wide range of surfaces, if only one technique is adopted, DART is preferable for the potential ACWA utilization scenarios, based on its lower dispersion of target species, utilization of a gas rather than a liquid as the working medium, and ability to efficiently ionize and detect trace levels of species in the gas phase. A DESI system with a cover shield to intercept dispersed contaminants may also be applicable. The use of instrument shields to minimize agent contamination would have to be investigated during instrument test and evaluation activities.

**Recommendation 4-2.** The Army should specify a list of requirements for an ambient ionization mass spectrometer system that would implement analytical capabilities specifically designed to respond to the challenges summarized in the different scenarios described in Chapter 3. Suggested mass analyzers are either triple quadrupole or linear ion trap, because both can be operated in selective reaction monitoring modes for validation of agent identification and reduction of false positives without requiring high mass resolution. The instrument's atmospheric pressure interface should be fitted with an optionally heated transfer line designed to serve as a multifunction sampling wand. Ionization can occur either at the end of the wand (desorbed surface species or ambient vapor) or at the atmospheric pressure sampling orifice of the mass spectrometer (vapor). In the first case, ions are transported, whereas in the second case, neutrals are generated and then transported to the ionizer. A system employing a wand should be tested for the efficacy of the analytical methodology to trace an expanding vapor plume back to its source. This would be especially beneficial in identifying and locating Type I and possibly Type II occluded spaces (see Box 3-3) as well as in identifying leaker munitions in both storage and processing areas. The overall ambient ionization mass spectrometric system should be portable, either cart-mounted or handheld, for maximum utility.

**Recommendation 4-3.** The sampling wand should accommodate a variety of sampling modes and interchangeable ion sources:

- A compact DART (and possibly DESI) source that can be mounted either at the end of the wand, with ions sampled through the wand, or directly on the mass spectrometer, with neutrals sampled through the wand.
- A thermal neutral desorption mode where hot gas is blown toward the surface and the desorbed neutrals are collected by a suction interface that directs them toward the ionization source attached to the mass spectrometer.
- A gas sampling mode where ambient air is drawn into the interface but no gas is blown to any surfaces.
- A sampling port that allows the user to manually wipe an area and either place the wipe directly on the sampling wand or insert it through the port into the plasma of the DART source attached to the mass spectrometer.
- A high-sensitivity air monitoring mode where a solid-phase microextraction fiber is exposed to the vapor to be sampled and directed into the plasma source.
- A high-sensitivity surface sampling mode where a polydimethylsiloxane membrane is attached to the surface to be sampled for an extended period of time and then exposed directly to the plasma for desorption ionization.
- A liquid sampling mode that allows manual sampling of a liquid pool or drip.

Ambient ionization mass spectrometry methods achieve their high level of sensitivity by utilizing highly efficient chemical ionization processes. Based on the committee's analyses of the ion chemistry of the chemical agents present in ACWA site weapons and published studies of mass spectrometric detection of these agents presented in Chapter 4, the committee finds:

**Finding 4-7.** The chemical agents VX, GB, and HD have markedly different acid-base properties in the gas phase. This is reflected in the quasimolecular ions observed and monitored when DART is employed for their detection. Only VX is observed as the protonated molecular ion, GB is detected as an adduct with ammonium ion, and HD is detected as the oxidized sulfoxide derivative, which has a proton affinity significantly higher than that of the parent molecule. While it has been demonstrated that all three species can be detected with high sensitivity in laboratory studies using the same instrumental conditions, it is not obvious that this multiagent detection capability will be possible when this experimental methodology is deployed and used in working environments. For example, ammonia in laboratories can originate from human breath and would likely not be as abundant in Class A environments where workers are in DPE suits and might have to be added to the sampling flow. In addition, the extent of sample oxidation in ambient ionization sources is known to be dependent on source operating conditions and environmental factors such as the relative humidity.

Chapter 3 presents several scenarios that describe potential uses of ambient ionization mass spectrometry that might expedite agent processing or closure activities. If ambient ionization mass spectrometry methods are to be used in these scenarios or other situations, they must be tested and evaluated prior to implementation. With this requirement in mind the committee finds and recommends:

**Finding 4-8.** MINICAMS are designed to monitor relatively large areas and alarm when levels exceed a specified limit. Although regarded as near-real-time detectors, they typically require 5 to 10 min for a single analysis. This is not particularly efficient when attempting to locate and define an area of agent contamination. For example, when exposure to agent is possible, MINICAMS sampling ports are moved over worker DPE suits to see if agent levels exceed 1 VSL, pausing in four quadrants to take a measurement. This is a time-consuming procedure.

**Recommendation 4-4.** Procedures developed and optimized in laboratory environments for the real-time detection of chemical agents using ambient ionization mass spectrometry should be verified in all working environments where they are likely to be deployed, using actual sample materials (e.g., activated charcoal from filter beds and worker masks, DPE suit material, and polymer-coated concrete).

**Recommendation 4-5.** Procedures should be developed for using ambient ionization mass spectrometry (e.g., DART, large-area DESI) to check worker DPE suits for contamination when workers are exiting Class A work areas. This approach could greatly reduce the time required for this activity, including verification of the effectiveness of decontamination procedures carried out prior to DPE suit removal when agent is detected.

The potential ambient ionization mass spectrometry application scenarios discussed in Chapters 3-5 rely on only relative detection of agent contamination on or in condensed materials. There is no requirement that absolute quantification of surface or bulk agent concentrations be determined. Currently, neither CMA nor ACWA has developed a health-based surface agent contamination hazard level (see Finding 2-2), and so clearing contaminated waste by showing that surface contamination is below a defined level is not currently feasible. ACWA is planning to clear primary liquid waste based on measuring bulk agent chemical concentrations using other analytical techniques. Reliable quantification of either surface or bulk agent contamination requires suitable calibration standards for any analytical technique, including ambient ionization mass spectrometry. Even if absolute agent surface contamination level measurements are not required, reliable bulk and surface agent contaminated standard materials are valuable to check ambient ionization mass spectrometry instrumental sensitivity and time response. Based on the discussions of these issues presented in Chapter 4, the committee finds and recommends:

**Finding 4-9.** Mass spectrometric detection methodology is able to provide relative quantification of analyte species, including chemical agents, over several orders of magnitude with excellent linearity between concentration and response for both gas- and liquid-phase analytes with appropriate sampling and ionization methods. Absolute quantification can be provided with appropriate reference standards. This is true of the ambient ionization mass spectrometric methods as well. However, quantifying amounts of target species on surfaces and adsorbed in solid substrates is more problematical.

**Recommendation 4-6.** The Army should develop reference standards to permit calibration of mass spectrometric instruments using DART (or other deployed ambient ionization sources) for analysis of chemical agents in gases and liquids. In the case of gas-phase samples, it would be useful to develop a reference standard that reliably provides a vapor-phase concentration equal to 1 vapor screening level of the target agent, both to quantify measurements and to verify acceptable performance during critical operations. Even though they may be less reliable for quantitative analysis, calibration standards and procedures should also be developed that ensure acceptable sensitivity for detection of trace amounts of agents on relevant surfaces.

To demonstrate the application of statistical data analysis techniques on a relevant ambient ionization data set, in Chapter 5 the committee presents analyses of DART measurement data for chemical agent concentrations in distilled water and 2-propanol (VX) and methylene chloride (GA, GB, and HD). These data were acquired at the Army's Edgewood Chemical Biological Center and are summarized in Nilles et al. (2009). Based on these analyses, the committee finds as follows:

**Finding 5-1.** New analytical methodologies (e.g., DART) have been demonstrated for relevant nerve agents (GB and VX) and blister agents (HD, HT, H) in simple liquid matrices (deionized water, isopropyl alcohol, methylene chloride). Similar data for relevant surfaces (e.g., metal, concrete, activated carbon, plastics, and iron oxide) are sparse.

**Finding 5-2.** Good precision and accuracy for DART techniques have been established for liquid matrices through the use of an internal standard. Challenges for development of internal standards for surface measurements include the presence of potential interferents;

heterogeneous solid matrix properties; and heterogeneous analyte distributions and difficulty in introducing any internal standards. Development of internal standards for more homogeneous liquid and gas-phase matrices is more straightforward. The development of an internal standard may or may not be practical for some surface analysis applications at the Pueblo Chemical Agent Destruction Pilot Plant or the Blue Grass Chemical Agent Destruction Pilot Plant. In the absence of an internal standard, the precision of the quantitative measurements may decrease.

Based on these findings, the committee recommends:

**Recommendation 5-1.** The use of DART, DESI, or related new analytical methodologies for surface area measurements at the Pueblo Chemical Agent Destruction Pilot Plant or the Blue Grass Chemical Agent Destruction Pilot Plant requires that the quality of measurements be determined and related calibration studies be performed for relevant matrices.

**Recommendation 5-2.** In the absence of an internal standard for surface measurements, the uncertainty in the measurement technologies (e.g., DART and DESI) should be established.

Based on its review early in Chapter 5 of the statistical analysis procedures presented in the ACWA *Chemical Agent Laboratory and Monitoring Quality Assurance Plan* (LMQAP) (U.S. Army, 2011b) and the role of valid upper confidence limits in compliance monitoring as discussed later in the chapter, the committee finds as follows:

**Finding 5-3.** The published protocols for statistical procedures for compliance monitoring by the Assembled Chemical Weapons Alternatives program (i.e., in the draft *Laboratory Monitoring Quality Assurance Plan*, Rev. 0, April 4, 2011) contain insufficient detail to provide guidance for compliance monitoring. The ambiguity of current publicly available documentation suggests that quantification of statistical variability has the potential to be inaccurate. Such inaccuracies may result in difficulties such as unnecessary destruction of uncontaminated waste and/or failure to identify contaminated waste.

Accordingly, the committee recommends:

**Recommendation 5-3.** The Assembled Chemical Weapons Alternatives program should reexamine existing protocols planned for compliance monitoring at PCAPP and BGCAPP and means for incorporating increased statistical rigor in the assessments to be performed.

The committee briefly addresses aspects of bias, precision, and detection limits of real-time analytical instrumentation, finding that computer software programs, when appropriately validated as recommended, can play a role in providing reliable real-time ambient ionization mass spectrometry measurements.

**Finding 5-4.** Reliable real-time computer programs able to interpret real-time chemical analyses enable instruments (generally using proprietary software) to convert intensity measurements to concentrations of agent and potential interferents.

**Recommendation 5-4.** Instrument software for use with ambient ionization mass spectrometry should be reviewed to ensure that it meets appropriate validation and verification criteria. This software should be tested by using simulated data to test different measurement scenarios (e.g., all data below detection limits, at detection limits, mixtures, hot spots, and so on).

Chapter 5 evaluations of statistical sampling issues for materials potentially contaminated by chemical agent led to the following finding and recommendation:

**Finding 5-5.** In some cases, analysis of direct surface and/or materials wipe sampling may complement or replace vapor screening level analysis, allowing more efficient and cost-effective closure operations.

**Recommendation 5-5.** If direct surface and/or materials wipe sampling analysis methods are adopted, appropriate statistical methods for characterizing the extent of contamination of surfaces, machinery, and/or materials should be employed.

Following the extensive discussion of statistical sampling plans in Chapter 5, the committee finds and recommends:

**Finding 5-6.** The use of statistical sampling will improve agent contamination detection and quantitation. For near-real-time measurement technologies, sequential sampling may be particularly valuable. Specific sampling plans will depend on the geometry of the contaminated area, contaminant spatial variability, and the goal of the measurement process.

**Finding 5-7.** The successful application of any measurement technology is a function not only of its capabilities, but also of the ultimate use of the data generated. The context and purpose of the measurements will determine the sampling scheme, precision, and accuracy required.

**Recommendation 5-6.** The Assembled Chemical Weapons Alternatives (ACWA) staff should have access to sufficient statistical expertise to develop effective sampling protocols for any application of ambient ionization monitoring. Once the resulting expert sampling protocols have been developed, ACWA headquarters monitoring staff or their contractors should then proceed to develop detailed standard operating procedures to guide monitoring technicians

## CONCLUSIONS

As reflected in the recommendations presented above, the committee concluded that ambient ionization mass spectrometry is a rapidly maturing and highly useful technology with specific available implementations capable of highly sensitive, real-time measurements of relative concentrations of chemical agents adsorbed on a variety of relevant surfaces and in some porous materials. Further, with suitable reference standards, absolute measurements of agent concentrations in ambient air and liquid solutions are feasible. If adopted, these capabilities might be very useful in supplementing the Army's traditional air and vapor headspace agent contamination measurements using current near-real-time agent monitors. A range of scenarios occurring during agent disposal operations and facility closure activities have been defined and developed by the committee to illustrate the potential utility of real-time ambient ionization mass spectrometric detection of chemical agent contamination.

Although commercially available ambient ionization mass spectrometry instrumentation in the specific configurations recommended by the committee may not currently be available off the shelf, the major components have been commercialized, and a number of analytical instrument vendors are capable of designing, assembling, and demonstrating instruments meeting potential ACWA specifications. Given the current schedules for anticipated PCAPP and BGCAPP weapons disposal (beginning in 2015 and 2020, respectively) and facility closure activities, it is very likely that these instruments could be specified, tested, and deployed quickly enough to be used at BGCAPP and PCAPP as suggested in this report.

In addition, as demonstrated by their work as reviewed in Chapters 4 and 5, Army scientists at the Edgewood Chemical and Biological Center, sited near ACWA headquarters, have significant experience in the application of ambient ionization mass spectrometric measurements of chemical agent concentrations and distributions and could be a resource for developing and testing specific ambient ionization technology implementations for ACWA.

Based on these considerations the committee's overarching finding and recommendation are as follows:

**Finding 6-1.** Suitably specified ambient ion mass spectrometry instrumentation could be utilized in a range of challenging activities at ACWA chemical weapons disposal facilities where real-time chemical agent contamination measurements may reduce the time and effort required to characterize chemical agent contamination of waste materials, process equipment, and work areas.

**Recommendation 6-1.** ACWA should carefully evaluate the capabilities of portable ambient ion mass spectrometry and its potential to provide faster and more accurate characterization of chemical agent contamination, as detailed in this report, and determine if these likely benefits justify the effort and investment required to specify, acquire, and deploy suitable implementations of this technology.

# References

Adamson, A. and A. Gast. 1997. Physical Chemistry of Surfaces (6th Edition.) New York, N.Y.: John Wiley & Sons.

Andrade, F., J. Shelley, W. Wetzel, M. Webb, G. Gamez, S. Ray, and G. Hieftje. 2008. Atmospheric pressure chemical ionization source. 1. Ionization of compounds in the gas phase. Analytical Chemistry 80(8): 2646-2653.

Anicich, V. 2003. JPL Publication 03-19. An index of the literature for bimolecular gas phase cation-molecule reaction kinetics. Pasadena, Calif.: Jet Propulsion Laboratory.

Badman, E. and R. Cooks. 2000. Miniature mass analyzers. Journal of Mass Spectrometry 35(6): 659-671.

Basile, F., S. Zhang, Y.-S. Shin, and B. Drolet. 2010. Atmospheric pressure-thermal desorption (AP-TD)/electrospray ionization-mass spectrometry for the rapid analysis of Bacillus spores. Analyst 135(4): 797-803.

Bizzigotti, G., H. Castelly, A. Hafez, W. Smith, and M. Whitmire. 2009. Parameters for evaluation of the fate, transport, and environmental impacts of chemical agents in marine environments. Chemical Reviews 109(1): 236-256.

Bohme, D. 1975. The kinetics and energetics of proton transfer. Pp 489-504 in Interactions Between Ions and Molecules, edited by P. Ausloos. New York, N.Y.: Plenum Press.

BPBGT (Bechtel Parsons Blue Grass Team). 2006. Operations and Closure Agent-Contaminated Waste Disposal Estimate Summary Report, BGCAPP WP-045, Revision 0, December 28. Richmond, Ky.: Bechtel Parsons Blue Grass Team.

BPT (Bechtel Pueblo Team). 2004. Initial Design for the Pueblo Chemical Agent Destruction Pilot Plant (PCAPP) Project, Rev. A—redacted for release to NRC, January 16. Aberdeen Proving Ground, Md.: Program Manager for Assembled Chemical Weapons Alternatives.

CDC (Centers for Disease Control and Prevention). 2010. Review of the Chemical Agent Air Monitoring Program at the Anniston, Pine Bluff, Tooele, and Umatilla Chemical Agent Disposal Facilities. Atlanta, Ga.: National Center for Environmental Health.

Chan, C.-C., M. Bolgar, S. Miller, and A. Attygalle. 2010. Desorption ionization by charge exchange (DICE) for sample analysis under ambient conditions by mass spectrometry. Journal of the American Society for Mass Spectrometry 21(9): 1554-1560.

Chaudhary, A., F. van Amerom, R. Short, and S. Bhansali. 2006. Fabrication and testing of a miniature cylindrical ion trap mass spectrometer constructed from low temperature co-fired ceramics. International Journal of Mass Spectrometry 251(1): 32-39.

Chen, H., N. Talaty, Z. Takats, and R. Cooks. 2005. Desorption electrospray ionization mass spectrometry for high-throughput analysis of pharmaceutical samples in the ambient environment. Analytical Chemistry 77(21): 6915-6927.

Chen, H., A. Wortmann, and R. Zenobi. 2007. Neutral desorption sampling coupled to extractive electrospray ionization mass spectrometry for rapid differentiation of biosamples by metabolomic fingerprinting. Journal of Mass Spectrometry 42(9): 1123-1135.

Chen, H., B. Hu, Y. Hu, Y. Huan, Z. Zhou, and X. Qiao. 2009. Neutral desorption using a sealed enclosure to sample explosives on human skin for rapid detection by EESI-MS. Journal of the American Society for Mass Spectrometry 20(4): 719-722.

Cheng, S.-C., T.-L. Cheng, H.-C. Chang, and J. Shiea. 2009. Using laser-induced acoustic desorption/electrospray ionization mass spectrometry to characterize small organic and large biological compounds in the solid state and in solution under ambient conditions. Analytical Chemistry 81(3): 868-874.

Chipuk, J. and J. Brodbelt. 2008. Transmission mode desorption electrospray ionization. Journal of the American Society for Mass Spectrometry 19(11): 1612-1620.

Cody, R. 2009. Observation of molecular ions and analysis of nonpolar compounds with the direct analysis in real time ion source. Analytical Chemistry 81(3): 1101-1107.

Cody, R., J. Laramée, and H. Durst. 2005. Versatile new ion source for the analysis of materials in open air under ambient conditions. Analytical Chemistry 77(8): 2297-2302.

Cooks, R., Z. Ouyang, A. Takats, and J. Wiseman. 2006. Ambient mass spectrometry. Science 311(5767): 1566-1570.

Cordell, R., K. Willis, K. Wyche, R. Blake, A. Ellis, and P. Monks. 2007. Detection of chemical weapon agents and simulants using chemical ionization reaction time-of-flight mass spectrometry. Analytical Chemistry 79(21): 8359-8366.

Costa, A. and R. Cooks. 2007. Simulation of atmospheric transport and droplet–thin film collisions in desorption electrospray ionization. Chemical Communications 38: 3915-3917.

Costa, A. and R. Cooks. 2008. Simulated splashes: Elucidating the mechanism of desorption electrospray ionization mass spectrometry. Chemical Physics Letters 464(1-3): 1-8.

Cotte-Rodriguez, I. and R. Cooks. 2006. Non-proximate detection of explosives and chemical warfare agent simulants by desorption electrospray ionization mass spectrometry. Chemical Communications 28: 2968-2970.

Cotte-Rodriguez, I., C. Mulligan, and R. Cooks. 2007. Non-proximate detection of small and large molecules by desorption electrospray ionization and desorption atmospheric pressure chemical ionization mass spectrometry: Instrumentation and applications in forensics, chemistry, and biology. Analytical Chemistry 79(18): 7069-7077.

Cotte-Rodriguez, I., Z. Takáts, N. Talaty, H. Chen, and R. Cooks. 2005. Desorption electrospray ionization of explosives on surfaces: Sensitivity and selectivity enhancement by reactive desorption electrospray ionization. Analytical Chemistry 77(21): 6755-6764.

Cressie, N. 1993. Statistics for Spatial Data, Revised Edition. Hoboken, N.J.: Wiley-Interscience.

D'Agostino, P. and C. Chenier. 2010. Desorption electrospray ionization mass spectrometric analysis of organophosphorus chemical warfare agents using ion mobility and tandem mass spectrometry. Rapid Communications in Mass Spectrometry 24(11): 1617-1624.

D'Agostino, P., J. Hancock, C. Chenier, and C. Jackson Lepage. 2006. Liquid chromatography electrospray tandem mass spectrometric and desorption electrospray ionization tandem mass spectrometric analysis of chemical warfare agents in office media typically collected during a forensic investigation. Journal of Chromatography A 1110(1-2): 86-94.

D'Agostino, P., C. Chenier, J. Hancock, and C. Jackson Lepage. 2007. Desorption electrospray ionisation mass spectrometric analysis of chemical warfare agents from solid-phase microextraction fibers. Rapid Communications in Mass Spectrometry 21(4): 543-549.

Dixon, R., M. Bereman, D. Muddiman, and A. Hawkridge. 2007. Remote mass spectrometric sampling of electrospray- and desorption electrospray-generated ions using an air ejector. Journal of the American Society for Mass Spectrometry 18(10): 1844-1847.

Dixon, R., J. Sampson, A. Hawkridge, and D. Muddiman. 2008. Ambient aerodynamic ionization source for remote analyte sampling and mass spectrometric analysis. Analytical Chemistry 80(13): 5266-5271.

Dixon, R., J. Sampson, and D. Muddiman. 2009. Generation of multiply charged peptides and proteins by radio frequency acoustic desorption and ionization for mass spectrometric detection. Journal of the American Society for Mass Spectrometry 20(4): 597-600.

Dzidic, I., D. Carroll, R. Stillwell, and E. Horning. 1975. Atmospheric pressure ionization (API) mass spectrometry: Formation of phenoxide ions from chlorinated aromatic compounds. Analytical Chemistry 47(8): 1308-1312.

Edison, S., L. Lin, B. Gamble, J. Wong, and K. Zhang. 2011. Surface swabbing technique for the rapid screening for pesticides using ambient pressure desorption ionization with high-resolution mass spectrometry. Rapid Communications in Mass Spectrometry 25(1): 127-139.

EPA (Environmental Protection Agency). 2007a. Method 3571 Extraction of Solid and Aqueous Samples for Chemical Agents. SW-846 Test Methods for Evaluating Solid Waste, Physical/Chemical Methods. Washington, D.C.: EPA Office of Solid Waste.

EPA. 2007b. Method 8271 Assay of Chemical Agents in Solid and Aqueous Samples by Gas Chromatograph/Mass Spectrometry, Electron Impact (GC/MS/EI). SW-846 Test Methods for Evaluating Solid Waste, Physical/Chemical Methods. Washington, D.C.: EPA Office of Solid Waste.

Fang, K.-T., D.K.J. Lin, P. Winker, and Y. Zhang. 2000. Uniform design: Theory and application. Technometrics 42(3): 237-248.

Ferguson, E. and F. Fehsenfeld. 1969. Water vapor ion cluster concentrations in the D Region. Journal of Geophysical Research 74(24): 5743-5751.

Galhena, A., G. Harris, L. Nyadong, K. Murray, and F. Fernandez. 2010. Small molecule ambient mass spectrometry imaging by infrared laser ablation metastable-induced chemical ionization. Analytical Chemistry 82(6): 2178-2181.

Gao, L., Q. Song, G. Patterson, R. Cooks, and Z. Ouyang. 2006. Handheld rectilinear ion trap mass spectrometer. Analytical Chemistry 78(17): 5994-6002.

Haapala, M., J. Pol, V. Saarela, V. Arvola, T. Kotiaho, R. Ketola, S. Franssila, T. Kauppila, and R. Kostiainen. 2007. Desorption atmospheric pressure photoionization. Analytical Chemistry 79(20): 7867-7872.

Haddad, R., R. Sparrapan, and M. Eberlin. 2006. Desorption sonic spray ionization for (high) voltage-free ambient mass spectrometry. Rapid Communications in Mass Spectrometry 20(19): 2901-2905.

Hagan, N., T. Cornish, R. Pilato, K. Van Houten, M. Antoine, T. Lippa, A. Becknell, and P. Demirev. 2008. Detection and identification of immobilized low-volatility organophosphates by desorption ionization mass spectrometry. International Journal of Mass Spectrometry 278(2-3): 158-165.

Harper, J., N. Charipar, C. Mulligan, X. Zhang, R. Cooks, and Z. Ouyang. 2008. Low-temperature plasma probe for ambient desorption ionization. Analytical Chemistry 80(23): 9097-9104.

Harris, G. and F. Fernandez. 2009. Simulations and experimental investigation of atmospheric transport in an ambient metastable-induced chemical ionization source. Analytical Chemistry 81(1): 322-329.

Harris, G., D. Hostetler, C. Hampton, and F. Fernandez. 2010. Comparison of the internal energy deposition of direct analysis in real time and electrospray ionization time-of-flight mass spectrometry. Journal of the American Society for Mass Spectrometry 21(5): 855-863.

Harris, G., A. Galhena, and F. Fernandez. 2011. Ambient sampling/ionization mass spectrometry: applications and current trends. Analytical Chemistry 83(12): 4508-4538.

Haunschmidt, M., C. Klampfl, W. Buchberger, and R. Hertsens. 2010. Determination of organic UV filters in water by stir bar sorptive extraction and direct analysis in real-time mass spectrometry. Analytical and Bioanalytical Chemistry 397(1): 269-275.

He, J., F. Tang, Z. Luo, Y. Chen, Z. Xu, R. Zhang, X. Wang, and Z. Abliz. 2011. Air flow assisted ionization for remote sampling of ambient mass spectrometry and its application. Rapid Communications in Mass Spectrometry 25(7): 843-850.

Helsel, D. 2004. Nondetects and Data Analysis: Statistics for Censored Environmental Data. Hoboken, N.J.: Wiley-Interscience.

Hiraoka, K., K. Nishidate, K. Mori, D. Asakawa, and S. Suzuki. 2007. Development of probe electrospray using a solid needle. Rapid Communications in Mass Spectrometry 21(18): 3139-3144.

Huang, M.-Z., C.-H. Yuan, S.-C. Cheng, Y.-T. Cho, and J. Shiea. 2010. Ambient ionization mass spectrometry. Annual Review of Analytical Chemistry 3: 43-65.

Johnson, M., L. Moore, and D. Ylvisaker. 1990. Minimax and maximin distance designs. Journal of Statistical Planning and Inference 26(2): 131-148.

Journel, A. and C. Huijbregts. 1978. Mining Geostatistics. Burlington, Mass.: Academic Press/Elsevier.

Karas, M. and F. Hillenkamp. 1988. Laser desorption ionization of proteins with molecular masses exceeding 10000 daltons. Analytical Chemistry 60(20): 2299-2301.

Kauppila, T., J. Wiseman, R. Ketola, T. Kotiaho, R. Cooks, and R. Kostiainen. 2006. Desorption electrospray ionization mass spectrometry for the analysis of pharmaceuticals and metabolites. Rapid Communications in Mass Spectrometry 20(3): 387-392.

Kertesz, V. and G. Van Berkel. 2010. Fully automated liquid extraction-based surface sampling and ionization using a chip-based robotic nanoelectrospray platform. Journal of Mass Spectrometry 45(3): 252-260.

Ketkar, S., S. Penn, and W. Fite. 1991a. Influence of coexisting analytes in atmospheric pressure ionization mass spectrometry. Analytical Chemistry 63(9): 924-925.

Ketkar, S., S. Penn, and W. Fite. 1991b. Real-time detection of parts per trillion levels of chemical warfare agents in ambient air using atmospheric pressure ionization tandem quadrupole mass spectrometry. Analytical Chemistry 63(5): 457-459.

Larame´e, J., H. Durst, T. Connell, and J. Nilles. 2008. Detectection of chemical warfare agents on surfaces relevant to homeland security by direct analysis in real-time spectrometry. American Laboratory 40(16): 16-20.

Lias, S., J. Bartmess, J. Liebman, J. Holmes, R. Levin, and W. Mallard. 1988. Gas-phase ion and neutral thermochemistry. Journal of Physical and Chemical Reference Data 17(Supplement 1): 1-861.

Linstrom, P. and W. Mallard, Eds. 2007. NIST Chemistry WebBook, NIST Reference Database No. 69. Gaithersburg, Md.: National Institute of Standards and Technology.

Ma, X., M. Zhao, Z. Lin, S. Zhang, C. Yang, and X. Zhang. 2008. Versatile platform employing desorption electrospray ionization mass spectrometry for high-throughput analysis. Analytical Chemistry 80(15): 6131-6136.

McGuire, J., C. Byers, S. Hulet, E. Jakubowski, and S. Thomson. 2008. A rapid and sensitive technique for assessing exposure to VX via GC–MS–MS analysis. Journal of Analytical Toxicology 32(1): 63-67.

Miao, Z. and H. Chen. 2009. Direct analysis of liquid samples by desorption electrospray ionization-mass spectrometry (DESI-MS). Journal of the American Society for Mass Spectrometry 20(1): 10-19.

Miao, Z., S. Wu, and H. Chen. 2010. The study of protein conformation in solution via direct sampling by desorption electrospray ionization mass spectrometry. Journal of the American Society for Mass Spectrometry 21(10): 1730-1736.

Miao, Z., H. Chen, P. Liu, and Y. Liu. 2011. Development of submillisecond time-resolved mass spectrometry using desorption electrospray ionization. Analytical Chemistry 83(11): 3994-3997.

Midey, A., T. Miller, and A. Viggiano. 2008. Kinetics of ion-molecule reactions with 2-chloroethyl ethyl sulfide at 298 K: A search for CIMS schemes for mustard gas. Journal of Physical Chemistry A 112(41): 10250-10256.

Midey, A., T. Miller, and A. Viggiano. 2009. Kinetics of ion-molecule reactions with Dimethyl Methylphosphonate at 298 K: Applicability for CIMS detection of GX. Journal of Physical Chemistry A 113(17): 4982-4989.

Midey, A., T. Miller, A. Viggiano, N. Bera, S. Maeda, and K. Morokuma. 2010. Ion chemistry of VX surrogates and ion energetics properties of VX: New suggestions for VX chemical ionization mass spectrometry detection. Analytical Chemistry 82(9): 3764-3771.

Mulligan, C., N. Talaty, and R. Cooks. 2006. Desorption electrospray ionization with a portable mass spectrometer: in situ analysis of ambient surfaces. Chemical Communications (16): 1709-1711.

Musselman, B. 2011. Open Air Desorption Ionization: Searching Surfaces for the Hidden Agenda, Gordon Research Conference on Detecting Illicit Substances: Explosives & Drugs. June 26-July 1, 2011, Lucca, Italy.

Na, N., M. Zhao, S. Zhang, C. Yang, and X. Zhang. 2007. Development of a dielectric barrier discharge ion source for ambient mass spectrometry. Journal of the American Society for Mass Spectrometry 18(10): 1859-1862.

Nefliu, M., A. Venter, and R. Cooks. 2006. Desorption electrospray ionization and electrosonic spray ionization for solid- and solution-phase analysis of industrial polymers. Chemical Communications (8): 888-890.

Nefliu, M., J. Smith, A. Venter, and R. Cooks. 2008. Internal energy distributions in desorption electrospray ionization (DESI). Journal of the American Society for Mass Spectrometry 19(3): 420-427.

Neidholdt, E. and J. Beauchamp. 2009. Ionization mechanism of the ambient pressure pyroelectric ion source (APPIS) and its applications to chemical nerve agent detection. Journal of the American Society for Mass Spectrometry 20(11): 2093-2099.

Neidholdt, E. and J. Beauchamp. 2011. Switched ferroelectric plasma ionizer (SwiFerr) for ambient mass spectrometry. Analytical Chemistry 83(1): 38-43.

Nemes, P. and A. Vertes. 2007. Laser ablation electrospray ionization for atmospheric pressure, in vivo, and imaging mass spectrometry. Analytical Chemistry 79(21): 8098-8106.

Nilles, J., T. Connell, and H. Durst. 2009. Quantitation of chemical warfare agents using the direct analysis in real time (DART) technique. Analytical Chemistry 81(16): 6744-6749.

Nilles, J., T. Connell, and H. Durst. 2010. Thermal separation to facilitate Direct Analysis in Real Time (DART) of mixtures. Analyst 135(5): 883-886.

NRC (National Research Council). 1994. Review of Monitoring Activities within the Chemical Stockpile Disposal Program. Washington, D.C.: National Academy Press.

NRC. 2002. Closure and Johnston Atoll Chemical Agent Disposal System. Washington, D.C.: The National Academies Press.

NRC. 2003. Acute Exposure Guideline Levels for Selected Airborne Chemicals: Volume 3. Washington, D.C.: The National Academies Press.

NRC. 2005a. Monitoring at Chemical Agent Disposal Facilities. Washington D.C.: The National Academies Press.

NRC. 2005b. Interim Design Assessment for the Pueblo Chemical Agent Destruction Pilot Plant. Washington, D.C.: The National Academies Press.

NRC. 2005c. Interim Design Assessment for the Blue Grass Chemical Agent Destruction Pilot Plant. Washington, D.C.: The National Academies Press.

NRC. 2005d. Impact of Revised Airborne Exposure Limits on Non-Stockpile Chemical Materiel Program Activities. Washington, D.C.: The National Academies Press.

NRC. 2006. Review of International Technologies for Destruction of Recovered Chemical Warfare Material. Washington, D.C.: The National Academies Press.

NRC. 2007. Review of Chemical Agent Secondary Waste Disposal and Regulatory Requirements. Washington, D.C.: The National Academies Press.

NRC. 2008a. Review of Secondary Waste Disposal Planning for the Blue Grass and Pueblo Chemical Agent Destruction Pilot Plants. Washington, D.C.: The National Academies Press.

NRC. 2008b. Review and Assessment of Developmental Issues Concerning the Metal Parts Treater Design for the Blue Grass Chemical Agent Destruction Pilot Plant. Washington, D.C.: The National Academies Press.

NRC. 2009a. Assessment of Explosive Destruction Technologies for Specific Munitions at the Blue Grass and Pueblo Chemical Agent Destruction Pilot Plants. Washington, D.C.: The National Academies Press.

NRC. 2009b. Disposal of Activated Carbon from Chemical Agent Disposal Facilities. Washington, D.C.: The National Academies Press.

NRC. 2010. Review of Closure Plans for the Baseline Incineration Chemical Agent Disposal Facilities. Washington, D.C.: The National Academies Press.

NRC. 2011. Assessment of Approaches for Using Process Safety Metrics at the Blue Grass and Pueblo Chemical Agent Destruction Pilot Plants. Washington, D.C.: The National Academies Press.

Nyadong, L., A. Galhena, and F. Fernandez. 2009. Desorption electrospray/metastable-induced ionization: A flexible multimode ambient ion generation technique. Analytical Chemistry 81(18): 7788-7794.

Perez, J., G. Harris, J. Chipuk, J. Brodbelt, M. Green, C. Hampton, and F. Fernandez. 2010. Transmission-mode direct analysis in real time and desorption electrospray ionization mass spectrometry of insecticide-treated bednets for malaria control. Analyst 135(4): 712-719.

PMACWA (Program Manager for Assembled Chemical Weapons Alternatives). 2006. Resource Conservation and Recovery Act Research, Development, and Demonstration Stage III, Class 3 Permit Modification Request for: The Department of the Army Assembled Chemical Weapons Alternatives Pueblo Chemical Agent Destruction Pilot Plant, Revision 0. Aberdeen Proving Ground, Md.: Program Manager for Assembled Chemical Weapons Alternatives. Available online at http://www.cdphe.state.co.us/hm/pcd/rdd/phase3mod/061130sec1.pdf. Last accessed November 16, 2011.

Rezenom, Y., J. Dong, and K. Murray. 2008. Infrared laser-assisted desorption electrospray ionization mass spectrometry. Analyst 133(2): 226-232.

Roach, P., J. Laskin, and A. Laskin. 2010. Molecular characterization of organic aerosols using nanospray-desorption/electrospray ionization-mass spectrometry. Analytical Chemistry 82(19): 7979-7986.

Rocke, D. and S. Lorenzato. 1995. A two-component model for measurement error in analytical chemistry. Technometrics 37(2): 176-184.

Rodriguez-Cruz, S. 2006. Rapid analysis of controlled substances using desorption electrospray ionization mass spectrometry. Rapid Communications in Mass Spectrometry 20(1): 53-60.

Sampson, J. and D. Muddiman. 2009. Atmospheric pressure infrared (10.6 micro m) laser desorption electrospray ionization (IR-LDESI) coupled to a LTQ Fourier transform ion cyclotron resonance mass spectrometer. Rapid Communications in Mass Spectrometry 23(13): 1989-1992.

Sampson, J., A. Hawkridge, and D. Muddiman. 2006. Generation and detection of multiply-charged peptides and proteins by matrix-assisted laser desorption electrospray ionization (MALDESI) Fourier transform ion cyclotron resonance mass spectrometry. Journal of the American Society for Mass Spectrometry 17 (12): 1712-1716.

Schäfer, K., J. Dénes, K. Albrecht, T. Szaniszló, J. Balog, R. Skoumal, M. Katona, M. Tóth, L. Balogh, and Z. Takáts. 2009. In vivo, in situ tissue analysis using rapid evaporative ionization mass spectrometry. Angewandte Chemie International Edition 48(44): 8240-8242.

Shiea, J., M.-Z. Huang, H.-J. Hsu, C.-Y. Lee, C.-H. Yuan, I. Beech, and J. Sunner. 2005. Electrospray-assisted laser desorption/ionization mass spectrometry for direct ambient analysis of solids. Rapid Communications in Mass Spectrometry 19(24): 3701-3704.

Smith, J., R. Flagan, and J. Beauchamp. 2002. Droplet evaporation and discharge dynamics in eectrospray ionization. Journal of Physical Chemistry A 106(42): 9957-9967.

Song, Y. and R. Cooks. 2007. Reactive desorption electrospray ionization for selective detection of the hydrolysis products of phosphonate esters. Journal of Mass Spectrometry 42(8): 1086-1092.

Song, L., A. Dykstra, H. Yao, and J. Bartmess. 2009a. Ionization mechanism of negative ion-direct analysis in real time: A comparative study with negative ion-atmospheric pressure photoionization. Journal of the American Society for Mass Spectrometry 20(1): 42-50.

Song, L., S. Gibson, D. Bhandari, K. Cook, and J. Bartmess. 2009b. Ionization mechanism of positive-ion direct analysis in real time: A transient microenvironment concept. Analytical Chemistry 81(24): 10080-10088.

Soparawalla, S., G. Salazar, R. Perry, M. Nicholas, and R. Cooks. 2009. Pharmaceutical cleaning validation using non-proximate large-area desorption electrospray ionization mass spectrometry. Rapid Communications in Mass Spectrometry 23(1): 131-137.

Steeb, J., A. Galhena, L. Nyadong, J. Janata, and F. Fernandez. 2009. Beta electron-assisted direct chemical ionization (BADCI) probe for ambient mass spectrometry. Chemical Communications(31): 4699-4701.

Stott, W., W. Davidson, and R. Sleeman. 1993. High throughput real time chemical contraband detection, Pp. 55 in Proceedings of the International Symposium on Substance Identification Technologies, Vol. 2092, edited by G. Harding, R.C. Lanza, I.J. Meyers, and P. A. Youngs. Innsbruck, Austria: SPIE.

Su, T. and W. Chesnavich. 1982. Parametrization of the ion-polar molecule collision rate constant by trajectory calculations. Journal of Chemical Physics 76: 5183-5185.

Takáts, Z., J. Wiseman, B. Gologan, and R. Cooks. 2004. Mass spectrometry sampling under ambient conditions with desorption electrospray ionization. Science 306(5695): 471-473.

Takáts, Z., J. Wiseman, and R. Cooks. 2005a. Ambient mass spectrometry using desorption electrospray ionization (DESI): Instrumentation, mechanisms and applications in forensics, chemistry,and biology. Journal of Mass Spectrometry 40(10): 1261-1275.

Takáts, Z., I. Cotte-Rodriguez, N. Talaty, H. Chen, and R. Cooks. 2005b. Direct, trace level detection of explosives on ambient surfaces by desorption electrospray ionization mass spectrometry. Chemical Communications(15): 1950-1952.

Talaty, N., Z. Takáts, and R. Cooks. 2005. Rapid in situ detection of alkaloids in plant tissue under ambient conditions using desorption electrospray ionization. Analyst 130(12): 1624-1633.

U.S. Army. 2008. DA-PAM 385-61 Toxic Chemical Agent Safety Standards. Washington, D.C.: Department of the Army.

U.S. Army. 2011a. USAE ACWA HQPL 5-12-0, U.S. Army Element, Assembled Chemical Weapons Alternatives (USAE ACWA) Chemical Agent Monitoring Concept Plan, Revision 0, April 4, 2011. Aberdeen Proving Ground, Md.: Assembled Chemical Weapons Alternatives.

U.S. Army. 2011b. Chemical Agent Laboratory and Monitoring Quality Assurance Plan, Revision 0. Aberdeen Proving Ground, Md.: U.S. Army Element, Assembled Chemicals Weapons Alternatives.

Van Berkel, G., M. Ford, and M. Deibel. 2005. Thin-layer chromatography and mass spectrometry coupled using desorption electrospray ionization. Analytical Chemistry 77(5): 1207-1215.

Van Berkel, G., S. Pasilis, and O. Ovchinnikova. 2008. Established and emerging atmospheric pressure surface sampling/ionization techniques for mass spectrometry. Journal of Mass Spectrometry 43(9): 1161-1180.

Vardeman, S. and J. Jobe. 1999. Statistical Quality Assurance Methods for Engineers. Hoboken, N.J.: Wiley.

Venter, A. and R. Cooks. 2007. Desorption electrospray ionization in a small pressure-tight enclosure. Analytical Chemistry 79(16): 6398-6403.

Verkouteren, J., J. Coleman, R. Fletcher, W. Smith, G. Klouda, and G. Gillen. 2008. A method to determine collection efficiency of particles by swipe sampling. Measurement Science and Technology 19(11): 1-12.

Wachs, T. and J. Henion. 2001. Electrospray device for coupling microscale separations and other miniaturized devices with electrospray mass spectrometry. Analytical Chemistry 73(3): 632-638.

Wang, H., W. Sun, J. Zhang, X. Yang, T. Lin, and L. Ding. 2010. Desorption corona beam ionization source for mass spectrometry. Analyst 135(4): 688-695.

Waybright, W. 2011. "PCAPP Air Monitoring Strategies and Secondary Waste Discussion," Briefing to ACWA Monitoring Committee and follow-up discussion on June 28, 2011, by Walter Waybright, PCAPP Laboratory Manager.

Weston, D., R. Bateman, I. Wilson, T. Wood, and C. Creaser. 2005. Direct analysis of pharmaceutical drug formulations using ion mobility spectrometry/quadrupole-time-of-flight mass spectrometry combined with desorption electrospray ionization. Analytical Chemistry 77(23): 7572-7580.

Williams, J. and J. Scrivens. 2005. Rapid accurate mass desorption electrospray ionisation tandem mass spectrometry of pharmaceutical samples. Rapid Communications in Mass Spectrometry 19(24): 3643-3650.

Wiseman, J., S. Puolitaival, Z. Takáts, R. Cooks, and R. Caprioli. 2005. Mass spectrometric profiling of intact biological tissue by using desorption electrospray ionization. Angewandte Chemie International Edition 44(43): 7094-7097.

Zhang, Y. and H. Chen. 2010. Detection of saccharides by reactive desorption electrospray ionization (DESI) using modified phenylboronic acids. International Journal of Mass Spectrometry 289(2-3): 98-107.

# Appendix A
# Biographical Sketches of Committee Members

**Charles E. Kolb**, *Chair*, is president and chief executive officer of Aerodyne Research, Inc., in Billerica, Massachusetts. His principle research interests have included atmospheric and environmental chemistry, combustion chemistry, and the chemical physics of rocket and aircraft exhaust plumes, and he has published over 200 archival journal articles and book chapters on these topics. He has served on several National Aeronautics and Space Administration panels dealing with environmental issues as well as on numerous NRC committees and boards dealing with atmospheric and environmental chemistry. Dr. Kolb also served on the NRC's Committee on Review and Evaluation of the Army Chemical Stockpile Disposal Program (member 1993-1998, vice-chair, 1998-2000), on the Committee on Review and Evaluation of Chemical Events at Army Chemical Demilitarization Facilities (chair, 2001-2002), and on the Committee on Monitoring at Chemical Agent Disposal Facilities (chair, 2004-2005), which examined ambient air monitoring at such facilities. He subsequently served on the NRC Standing Committee on Chemical Demilitarization (2005-2010). Dr. Kolb has also been appointed a national associate of the National Academies. He is a fellow of the American Physical Society, the American Chemical Society, the American Geophysical Union, the American Association for the Advancement of Science, and the Optical Society of America. Dr. Kolb graduated from the Massachusetts Institute of Technology with a B.S. in chemical physics and from Princeton University with an M.A. and a Ph.D. in physical chemistry.

**Jesse L. Beauchamp**, NAS, is Mary and Charles Ferkel Professor of Chemistry at the California Institute of Technology. Beginning with the first development of ion cyclotron resonance spectroscopy (Ph.D. thesis, Harvard, 1967), Professor Beauchamp has been involved with the development and application of mass spectrometry and other spectroscopic methods to a wide range of scientific investigations over the past 46 years. Major scientific contributions have involved the development of new instruments and techniques for studies of the structures, reaction dynamics, and properties of organic, inorganic, and biological molecules and ions in the gas phase. Current research efforts include development of novel reagents and methods for proteomics (e.g., cross-linking reagents and stable isotope labels), studies of the detailed mechanism of electron capture and electron transfer dissociation, investigations of reaction dynamics at gas-liquid interfaces, studies of the structure and reaction dynamics of small molecular clusters, development of instrumentation for in situ elemental and chemical analysis on Mars, and an examination of the chemistry of Titan's atmosphere and surface. Dr. Beauchamp was elected to the National Academy of Sciences in 1981. In addition to major awards from the American Chemical Society in 1981, 1999, and 2003, in 2007 he received the Distinguished Contribution Award from the American Society for Mass Spectrometry. Dr. Beauchamp received a B.S. in 1964 from the California Institute of Technology and a Ph.D. in 1967 from Harvard University.

**Robert A. Beaudet** recently retired from the faculty of the University of Southern California. He received his Ph.D. in physical chemistry from Harvard University in 1962. From 1961 to 1963, he was a U.S. Army officer in the Chemical Branch and served at the Jet Propulsion Laboratory as a research scientist. He joined the faculty of the University of Southern California in 1962 and served continuously in the Department of Chemistry from that time. Most of his academic career has been devoted to research in molecular structure and molecular spectroscopy. He also has served on Department of Defense committees addressing both offensive and defensive aspects surrounding chemical and biological warfare agents. He was chair of an Army Science Board committee that addressed chemical detection and trace gas analysis. Dr. Beaudet served as a member of the NRC's Board on Army Science and Technology (BAST), as a member of the NRC Committee on Review of the Non-Stockpile Chemical Materiel Disposal Program, and as a BAST liaison to the Committee on Review and Evaluation of the Army Chemical Stockpile Disposal Program (Stockpile Committee). He was also the chair of an Air Force technical conference on chemical warfare decontamination and protection. Dr. Beaudet has participated in numerous studies by the National Research Council (NRC) dealing with chemical and biological sensor technologies and properties and detection of energetic materials. Over the past decade, he has chaired or served as a member on numerous NRC committees examining issues on the design of the Assembled Chemical Weapons Alternatives program pilot plant facilities in Colorado and Kentucky.

**Joan B. Berkowitz** is currently managing director of Farkas Berkowitz and Company. She graduated from the University of Illinois with a Ph.D. in physical chemistry. Dr. Berkowitz has extensive experience in the area of environmental and hazardous waste management, a knowledge of available technologies for the cleanup of contaminated soils and groundwater, and a background in physical and electrochemistry. She has contributed to several EPA studies, been a consultant on remediation techniques, and assessed various destruction technologies. Dr. Berkowitz has written numerous publications on hazardous waste treatment and environmental subjects. She is an adjunct professor in the Graduate School of Management and Technology at the University of Maryland, University College, and winner of the 2004 Drazek Award for excellence in teaching. Dr. Berkowitz has for more than a decade been a participant on NRC studies concerning the chemical stockpile disposal program generally, and including the ACWA program for developing alternative disposal methods to incineration specifically. Among the 10 NRC studies on chemical demilitarization in which Dr. Berkowitz has participated since 1995 have been studies of the Committee to Review and Assess Developmental Issues Concerning the Metal Parts Treater Design for the Blue Grass Chemical Agent

Destruction Pilot Plant and the Committee to Examine the Disposal of Activated Carbon from the Heating, Ventilation and Air Conditioning Systems at Chemical Agent Disposal Facilities. She is also currently a member of the standing Committee on Chemical Demilitarization.

**Hao Chen** is an assistant professor in the Department of Chemistry and Biochemistry of Ohio University. His research in the chemical and life sciences in the area of organic and biological mass spectrometry focuses on how mass spectrometry can be applied at the interface between chemistry, physical organic chemistry, biology, and materials science. Examples include ion chemistry using ambient mass spectrometry for novel applications in bioanalytical chemistry. This involves developing methodologies for selective biomolecule detection in complex matrices. Other examples are analytical applications of ambient ion dissociation in proteomics and the chemical footprinting of proteins. Dr. Chen has authored or coauthored extensively in these areas since receiving his Ph.D. from Purdue University in 2005. He has also been a reviewer for a number of professional journals on analytical chemistry, mass spectrometry, and proteome research.

**Adrienne T. Cooper** is an associate professor of biological and agricultural systems engineering at the Florida Agricultural and Mechanical University. Previously, she was an assistant professor in the Department of Civil and Environmental Engineering at Temple University. She has 20 years of experience in chemical and environmental engineering, including process engineering, process and waste treatment development, and environmental regulation. Dr. Cooper conducts research in catalytic processes for environmental treatment and remediation and pollution prevention. She is a recipient of the National Science Foundation's CAREER award for her research on the development of photochemical reactors for water treatment and remediation. She has authored numerous publications and made presentations in her field. Dr. Cooper has served as a member of several National Research Council committees on issues pertaining to the disposal of stockpiled chemical agents and munitions and recovered (nonstockpile) chemical warfare materiel since 1999, including the 2009 report *Assessment of Explosive Destruction Technologies for Specific Munitions at the Blue Grass and Pueblo Chemical Agent Disposal Pilot Plants*. She holds a Ph.D. in environmental engineering from the University of Florida and a B.S. in chemical engineering from the University of Tennessee.

**Facundo M. Fernández** is an associate professor in the School of Chemistry and Biochemistry at the Georgia Institute of Technology. He was born in Buenos Aires, Argentina, and received his M.Sc. in chemistry from the College of Exact and Natural Sciences at Buenos Aires University in 1995 and his Ph.D. in analytical chemistry from the same University in 1999. In August 2000, he joined the research group of Richard Zare in the Department of Chemistry at Stanford University. His work focused on several aspects of Hadamard transform time-of-flight mass spectrometry with an emphasis on coupling this technique to capillary-format separation methods. In 2002, he joined the group of Vicki Wysocki in the Department of Chemistry at the University of Arizona, which was developing novel tandem mass spectrometers. In 2004 he joined the School of Chemistry and Biochemistry at the Georgia Institute of Technology. He is the author of over 70 peer-reviewed publications and numerous presentations at international conferences. He has received various awards, including the NSF CAREER award, the CETL/BP Teaching award, and the Ron A. Hites best paper award from the American Society for Mass Spectrometry, among others. His current research interests include the study and development of new methods and instrumentation in analytical mass spectrometry for forensics, metabolomics, and imaging.

**Robert D. Gibbons,** IOM, is director of the Center for Health Statistics and professor of medicine, health studies, and psychiatry at the University of Chicago. He received his doctorate in statistics and psychometrics from the University of Chicago in 1981. He spent the first 30 years of his career at the University of Illinois at Chicago (UIC) (1981-2010), where he directed the Center for Health Statistics, a consortium of 15 statisticians working in both theoretical and applied areas of environmetrics, chemometrics, biometrics, and psychometrics. Support for his research comes from numerous grants and contracts from the NIH, NIMH, ONR, NCI, and the MacArthur Foundation. Recognition for his work includes a Young Scientist Award from the Office of Naval Research, a Research Scientist Award from NIH, the Harvard Award for lifetime contributions to psychiatric epidemiology and biostatistics, the Lucaks award for contributions to environmental statistics in the twentieth century, and two Youden prizes (2001 and 2006) from the American Statistical Association for statistical contributions to the field of chemistry. Dr. Gibbons is a fellow of the American Statistical Association and a member of the Institute of Medicine of the National Academies. He has authored more than 200 peer-reviewed scientific papers and five books. Most recently Dr. Gibbons received the 2009 Outstanding Statistical Application Award from the American Statistical Association and the 2009 Distinguished Faculty Award from the University of Illinois at Chicago. In 2010 Dr. Gibbons retired from UIC and joined the faculty of the University of Chicago.

**John A. McLean** is an assistant professor in the Department of Chemistry at Vanderbilt University and a faculty fellow in the Institute of Chemical Biology and the Institute of Integrative Biosystems Research and Education. His recent awards include an Excellence in Teaching award from the student members of the American Chemical Society, a Defense Threat Reduction Agency research award, an American Society for Mass Spectrometry research award, a Spectroscopy Society of Pittsburgh award, an R&D 100 Award, and the Bunsen-Kirchhoff Prize from the Gesellschaft Deutscher Chemiker. His research interests focus on the design, conceptualization, construction, and application of technologies for structural mass spectrometry, in particular for studies in structural proteomics, systems biology, biophysics, and bionanotechnology. Dr. McLean received a B.S. in chemistry from the University of Michigan and in 2001 received a Ph.D. in chemistry from George Washington University.

**Max D. Morris** is professor of statistics and industrial engineering at Iowa State University. His research addresses the statistical design of experiments, the theory and application of linear statistical models, and the theory and application of Gaussian process models. Recent areas of active research have included factor screening designs, design of spatial sampling plans, and pattern matching algorithms for forensic applications. He has done collaborative work with the scientific staff of Ames Laboratory. Dr. Morris was elected as a fellow of the American Statistical Association in 1994. In 2002, he received the Jerome Sacks Award for Cross-Disciplinary Research from the National Institute of Statistical Sciences. Dr. Morris received a B.S. in mathematics from Oklahoma State University in 1973 and an M.S. (1975) and a Ph.D. (1977) in statistics from Virginia Polytechnic Institute and State University.

**Donald W. Murphy**, NAE, now retired from Bell Laboratories, Lucent Technologies, was director of the Applied Materials Research Department. He has since been a visiting researcher in the Chemistry Department of the University of California at Davis and an independent consultant. Dr. Murphy's research interests center on the synthesis of inorganic materials and on energy storage and conversion. A fellow of the American Association for the Advancement of Science and a member of the American Chemical Society, the American Physical Society, and the National Academy of Engineering, Dr. Murphy has also published widely in his field. Dr. Murphy acquired familiarity with chemical agent demilitarization operations through prior service as a member of several NRC committees, including the committee that wrote the 2005 report *Monitoring at Chemical Agent Disposal Facilities*, which assessed the status and opportunities concerning ambient air monitoring at chemical demilitarization facilities. Dr. Murphy has a B.S. in chemistry from Harvey Mudd College and a Ph.D. in inorganic chemistry from Stanford University.

**C. Shane Reese** is professor of statistics and associate chair at Brigham Young University. He received his Ph.D. in statistics from Texas A&M University. He is associate editor of the *Journal of the American Statistical Association*. He also serves as chair of the association Council of Chapters Governing Board and as a member of the ASA Committee on Science and Public Affairs. He has also served as president of the Albuquerque and Utah chapters of the association. At Brigham Young he is a member of the faculty advisory committee (faculty senate) and university athletic advisory committee. Dr. Reese's research interests include Bayesian hierarchical models, Bayesian design and analysis of computer experiments, and Bayesian reliability and sports statistics. He has served as a member of the National Academy of Sciences Committee on Biological Standoff Detection Systems.

**Lorenz R. Rhomberg** is a principal at Gradient and an expert in quantitative risk assessment, including dose-response analysis, pharmacokinetic modeling, and probabilistic methods, with special experience in chlorinated solvents and endocrine-active agents. His experience encompasses work relating to a variety of regulatory programs, including CERCLA, FIFRA, and TSCA. Before joining Gradient, Dr. Rhomberg was on the faculty of the Harvard School of Public Health and was employed by the Environmental Protection Agency. Dr. Rhomberg has been involved in policy development with a focus on current issues in the interpretation of toxicological data in human health risk assessment. He was recognized as Outstanding Practitioner of the Year by the Society for Risk Analysis in 2009. Dr. Rhomberg has a B.Sc. (with honors) in biology from Queen's University in Ontario, Canada, and a Ph.D. in population biology from the State University of New York at Stony Brook.

**Albert A. Viggiano**, a research chemist with the Space Vehicles Directorate of the Air Force Research Laboratory, graduated from the University of California at Berkeley with a B.Sc. in chemistry (with highest honors) in 1976. He received his Ph.D. in chemical physics from the University of Colorado at Boulder in 1980. Since graduate school, he has been active in the fields of atmospheric ion chemistry and kinetics, specifically in measuring ion molecule reaction rates of interest to atmospheric chemistry. Dr. Viggiano developed a chemical ionization mass spectrometer detection scheme for studying the thermal decomposition of $N_2O_5$ and was involved in the chemistry research that allowed sulfuric acid measurements in the atmosphere to be made. As a postdoctoral fellow in Heidelberg, Germany, he used these measurements to derive the first height profiles of sulfuric acid in the atmosphere and was involved in obtaining and analyzing in situ mass spectrometric data on the ion composition of the stratosphere. Since coming to the Air Force Research Laboratory in 1983 he has worked on measurements of ion kinetics related to a number of problems over a broad range of conditions. Dr. Viggiano has developed a technique that allows the effects of internal energy on the reactivity of gas-phase ion–molecule reactions to be studied. Using this technique, he has studied the effect of rotational energy on reactivity in more systems than have been studied anywhere. He has been instrumental in developing a technique to measure ion–molecule reactions at temperatures over 1000 K for the first time. The addition of a supersonic expansion source to the SIFDT allows for the measurement of mass selected cluster ions at thermal energy for the first time. A high-pressure turbulent flow tube is the first of its type for studying ion–molecule reactions. Most recently, he has developed a technique that allows for measuring the kinetics of numerous unique plasma reactions, including a heretofore previously undiscovered process, electron-catalyzed mutual neutralization. Dr. Viggiano has authored or coauthored over 320 papers and book chapters and has given 80 seminars at universities and laboratories and over 300 presentations at scientific meetings. He was lead author of the paper of the year at the Phillips Laboratory in 1993 and spent 2 months at the Max Planck Institut für Kernphysik working with Frank Arnold under the Air Force Window on Europe program. He won the Loeser award in 1997 and has won the Air Force Basic Research Award and the Air Force Science and Technology Achievement Award.

# Appendix B
# Committee Meetings

## FIRST MEETING
## FEBRUARY 22-24, 2011, ABERDEEN, MARYLAND

*Objectives*

- Conduct National Research Council (NRC) introduction to study procedure (administrative actions, committee introductions, composition/balance/bias discussions, portal usage, etc.)
- Discuss approach to Statement of Task, including background, and confirm scope during a review with sponsor
- Receive detailed briefings from ACWA staff and others
- Refine working report outline and review report writing process
- Appoint chapter leads and teams and make initial assignments
- Develop plan for future meeting milestones

*Briefings*

*ACWA Program Overview,* James Richmond, Director, Risk Management, U.S. Army Element, Assembled Chemical Weapons Alternatives

*PCAPP Site Project Overview,* Scott Susman, PCAPP Systems Engineer, U.S. Army Element, Assembled Chemical Weapons Alternatives

*BGCAPP Site Project Overview,* Darren Dalton, BGCAPP Systems Engineer, U.S. Army Element, Assembled Chemical Weapons Alternatives

*Relevant Health and Safety Standards,* Peter Spaeth, Safety Engineer, U.S. Army Element, Assembled Chemical Weapons Alternatives

*Relevant Environmental Standards,* Vicki Strause, Environmental Scientist, and Jon Ware, Environmental Scientist, U.S. Army Element, Assembled Chemical Weapons Alternatives

*Sources and Amounts of Agent-Contaminated Wastes,* Gary S. Groenewold, Idaho National Laboratory, former member of NRC ACWA Secondary Waste Committee; current member of NRC Committee on Chemical Demilitarization

*Current Monitoring Approaches in the Chemical Demilitarization Program,* Jeff Kiley, Chief, Quality Assurance, U.S. Army Chemical Materials Agency

*U.S. Army Technology Research in Low-Level Contaminant Monitoring Capabilities,* H. Dupont Durst, Research Directorate, U.S. Army Edgewood Chemical Biological Center

*General Status Review on Low-Level Contaminant Monitoring Capabilities,* R. Graham Cooks, Department of Chemistry, Purdue University

*Agent Fate Program and Other Relevant Work,* H. Dupont Durst, Research Directorate, U.S. Army Edgewood Chemical Biological Center

## SECOND MEETING
## JUNE 28-30, 2011, PUEBLO, COLORADO

*Objectives*

- Conduct NRC administrative actions, including composition and balance discussion for new member, and review NRC policies on intellectual property attributions
- Tour PCAPP site with emphasis on monitoring systems and procedures
- Obtain relevant information from site specialists and ACWA staff
- Review and refine draft text inputs for report
- Refine working report outline and reassess conformance with statement of task
- Identify additional information gathering needed
- Plan future activities, meetings, and text development deadlines

*Briefings*

*Update on ACWA Program Developments,* Carl Anderson, ACWA Safety & Surety Engineer

*PCAPP Site Orientation,* Video presentation

*Monitoring Strategies,* Walter Waybright, PCAPP Laboratory Manager (Battelle)

*Tour of Pueblo Chemical Agent Destruction Pilot Plant*

*Open Discussion and Question and Answer Session*

# THIRD MEETING
## AUGUST 15-16, 2011, KECK CENTER, WASHINGTON, D.C.

*Objectives*

- Conduct committee discussions on constructing and coordinating the text of each chapter toward conceptually realistic situations relating to the potential applicability and utility of using new complementary monitoring methods at BGCAPP and PCAPP
- Review and discuss information received from sponsor
- Identify additional information gathering to be pursued
- Determine strategy for site visit to Blue Grass Army Depot
- Perform page-by-page review of current draft text for each chapter
- Develop and/or refine draft findings and recommendations along with necessary supporting text
- Determine dates and location for next full committee meeting
- Review NRC policies on intellectual property attributions

*Participants in Teleconference with ACWA Personnel:*

James Richmond, Director, Risk Management, U.S. Army Element, Assembled Chemical Weapons Alternatives

Debra A. Michaels, Chemical Operations Officer, PCAPP Field Office

Anthony Reed, Deputy Site Project Manager for BGCAPP

John Coyne, SAIC, Risk Management/Chemistry/Science and Technology

Susan Ankrom, SAIC, ACWA Deputy Program Manager Engineering

Beth Haldane, SAIC, Chemist

John Barton, Battelle, Chief Scientist

Tommy Adams, Battelle, Laboratory Manager

# FOURTH MEETING
# BECKMAN CENTER, IRVINE, CALIFORNIA

*Objectives*

- Conduct committee discussions on how well text of each chapter addresses developed scenarios relating to the potential applicability and utility of using new, complementary monitoring methods at BGCAPP and PCAPP
- Review and discuss recent information received from sponsor
- Perform page-by-page review of text for each chapter
- Develop and/or refine findings and recommendations along with necessary supporting text
- Produce preconcurrence draft

*Briefings*

*ACWA Monitoring Update,* Jeff Kiley, Chief, Quality Assurance, U.S. Army Element, Assembled Chemical Weapons Alternatives

# Appendix C
# Commercial Sources of Ambient Ionization Mass Spectrometry Instrumentation

## OVERVIEW

While commercially available ambient ionization mass spectrometry instrumentation in the specific configurations recommended by the committee in Chapter 4 may not be currently available off the shelf, the major components have been commercialized and a number of analytical instrument vendors are capable of designing, assembling, and demonstrating instruments meeting potential ACWA specifications. A concise summary of commercial analytical instrument companies producing ambient ionization mass spectrometry instrumentation is presented below. This list should be considered indicative rather than exhaustive and should not be considered an endorsement by either the committee or the National Research Council.

## AMBIENT IONIZATION MASS SPECTROMETRY TECHNOLOGIES

### Direct Analysis in Real Time (DART)

*JEOL USA, Inc.*
*11 Dearborn Road, Peabody, Massachusetts 01960; 978-535-5900*

salesinfo@jeol.com
www.jeolusa.com/HOME/tabid/174/Default.aspx

U.S. subsidiary of JOEL Ltd. Japan; core owner of DART patents. Its key product is AccuTOF DART, which couples the DART ion source with the high-resolution, accurate mass capability of the AccuTOF time-of-flight mass spectrometer.

*IonSense, Inc.*
*999 Broadway, Saugas, Massachusetts 01906; 781-484-1043*

Info@IonSense.com
www.ionsense.com

U.S. licensee of JOEL DART patents. Key products are DART ionization sources, DART SVP controllers, DART-related experimental modules. Also offers atmospheric solids analysis probe (ASAP).

*Waters Corporation*
*34 Maple Street, Milford, Massachusetts 01757; 800-252-4752*

customerservice@waters.com
www.waters.com

The company's key products are mass spectrometers and integrated atmospheric solids analysis probes (ASAP).

### Desorption Electro Spray Ionization (DESI)

*Prosolia, Inc.*
*351 West 10th Street, Indianapolis, Indiana 46202; 317-278-6171*

info@prosolia.com
www.prosolia.com

The key products are Omni Spray (DESI) ion sources compatible with a range of commercial mass spectrometers.

### Laser Ablation Electrospray Ionization (LAESI)

*Protea Biosciences, Inc.*
*955 Hartman Run Road, Morgantown, West Virginia 26507; 877-776-8321*

stephen.turner@proteabio.com
www.proteabio.com

The key products are LAESI DP-1000 ionization probes.

## Liquid Extraction Surface Analysis (LESA)

*Advion*
*10 Brown Road, Ithaca, NewYork 14850; 607-266-9162*

info@advion.com
www.advion.com/biosystems/triversa-nanomate/LESA/

Advion's key product is LESA nano-electrospray ionization source compatible with a range of commercial mass spectrometers.

# Appendix D
# Statistical Calibration

In the following, the ordinary least squares (OLS) and the weighted least squares (WLS) approaches to estimating the calibration function and related interval are reviewed.

## OLS ESTIMATION[1]

As preparation for the following discussion, consider the relationship between response signal $y$ and spiking concentration $x$ in the region of the detection and quantification limits as a linear function of the form

$$y = \beta_0 + \beta_1 x + \varepsilon \tag{1}$$

where $\varepsilon$ is a random variable that describes the deviations from the regression line, distributed with mean 0 and constant variance $\sigma_{y \cdot x}^2$. The assumption of constant variance is not critical to this approach and will be relaxed in a later section; however, it is useful to simplify the initial exposition. The sample regression coefficient

$$b_1 = \frac{\sum_{i=1}^{n}\left[(x_i - \bar{x})y_i\right]}{\sum_{i=1}^{n}\left[(x_i - \bar{x})^2\right]} \tag{2}$$

provides an estimate of the population parameter $\beta_1$ (i.e., the slope of the calibration function). The sample intercept

$$b_0 = \bar{y} - b_1 \bar{x} \tag{3}$$

provides an estimate of the population parameter $\beta_0$ (i.e., the intercept of the calibration function, which describes the mean instrument response or measured concentration when

---

[1] This section is adapted from Gibbons, 1995.

true concentration $x = 0$). An unbiased sample estimate of $\sigma_{y \cdot x}^2$ (i.e., the variance of deviations from the population regression line) is given by

$$s_{y \cdot x}^2 = \sum_{i=1}^{n}(y_i - \hat{y}_i)^2/(n-2) \qquad (4)$$

where $\hat{y}_i = b_0 + b_1 x_i$.

## WLS ESTIMATION[2]

When variance is not constant, as is typically the case in the calibration setting, then the previous OLS solution for constant or "homoscedastic" errors no longer applies. There are several approaches to this problem, but in general the most widely accepted approach is to model the variance as a function of true concentration $x$ and to then use the estimated variances as weights in estimating the calibration parameters, which are now denoted as $\beta_{0w}$ and $\beta_{1w}$.

The weighted least squares regression of measured concentration or instrument response ($y$) on true concentration ($x$) is denoted by

$$\hat{y}_{wi} = b_{0w} + b_{1w} x_i \qquad (5)$$

where

$$b_{1w} = \frac{\sum_{i=1}^{n}[(x_i - \bar{x}_w)y_i/k_i]}{\sum_{i=1}^{n}[(x_i - \bar{x}_w)^2/k_i]} \qquad (6)$$

$$b_{0w} = \bar{y}_w - b_{1w}\bar{x}_w \qquad (7)$$

$$\bar{y}_w = \frac{\sum_{i=1}^{n}[y_i/k_i]}{\sum_{i=1}^{n}[1/k_i]} \qquad (8)$$

---

[2]This section is adapted from Gibbons and Bhaumik, 2001.

$$\bar{x}_w = \frac{\sum_{i=1}^{n}[x_i/k_i]}{\sum_{i=1}^{n}[1/k_i]} \qquad (9)$$

and the weight $k_i = s_{x_i}^2$ is the variance for sample $i$, which is computed from those samples with true concentration $x_i = x$. The weighted residual variance is

$$s_w^2 = \sum_{i=1}^{n}[(y_i - \hat{y}_{wi})^2/k_i]/(n-2) \qquad (10)$$

## ESTIMATING THE WEIGHTS[3]

When the number of replicates at each concentration is small, as is typically the case, or there are no replicates, the observed variance at each concentration provides a poor estimate of the true population variance. Two better alternatives are to (1) model the observed variance or standard deviation as a function of true concentration or (2) model the sum of squared residuals as a function of concentration. The latter approach can also be performed iteratively, in which improved estimates of $\beta_0$ and $\beta_1$ are obtained from weights computed from the current sum of squared residuals on each iteration. These new estimates of $\beta_0$ and $\beta_1$ are in turn used to obtain a new set of estimated weights and so forth until convergence. This algorithm is commonly termed "iteratively reweighted least squares." An essential element of either approach is to identify a plausible model for the variance function. The following sections consider a few models that are particularly well suited to this problem.

### Rocke and Lorenzato Model

To measure the true concentration of an analyte $(x)$, the traditional simple linear calibration model, $y = \beta_0 + \beta_1 x + e$ with the standard normality assumption on errors, is not appropriate, as it fails to explain increasing measurement variation with increasing analyte concentration, which is commonly observed in analytical data. To overcome this situation, one may propose a log linear model, for example, $y = xe^{\eta}$, where $\eta$ is a normal variable with mean $0$ and standard deviation $\sigma_{\eta}$. This model also fails to explain near-constant measurement variation of $y$ for low true concentration level $x$ (Rocke and Lorenzato, 1995). To better model the calibration curve, Rocke and Lorenzato (1995) proposed a combined model that has both types of errors:

---

[3]This section is adapted from Gibbons and Bhaumik, 2001.

$$y_{jr} = \beta_0 + \beta_1 x_{jr} e^{\eta} + e_{jr} \tag{11}$$

where $y$ is the $r$th measurement at the $j$th concentration level, $x_{jr}$ is the corresponding true concentration, and $\beta_0$ and $\beta_1$ are the fixed calibration parameters. In this model, $\eta$ represents proportional error at higher true concentrations and the $e_{jr}$'s are the additive errors that are present primarily at low concentrations. Now assume that $\eta$ and the $e_{jr}$'s are independent and follow normal distributions with means 0's and variances $\sigma_\eta^2$ and $\sigma_e^2$, respectively. Data near zero (i.e., $x; 0$) determine $\sigma_e^2$, and data for large concentrations determine $\sigma_\eta^2$. The model specification also indicates that errors at larger concentrations are lognormally distributed and at low concentrations are normally distributed, which agrees with common experience.

In their original paper, Rocke and Lorenzato (1995) derived the maximum likelihood estimators for their model based on maximizing the likelihood function:

$$\prod_{i=1}^{n} \frac{1}{2\pi \sigma_e \sigma_\eta} \int_{-\infty}^{\infty} e^{\frac{-\eta^2}{2\sigma_\eta^2}} e^{-\frac{(y_i - \beta_0 - \beta_1 x_i e^\eta)^2}{2\sigma_e^2}} d\eta \tag{12}$$

These computations require complex numerical evaluation of the required integrals. Alternatively, Gibbons et al. (1997) and Rocke and Durbin (1998) have described a WLS solution that involves the following algorithm:

1. Use OLS regression to find initial estimates of $\beta_0$ and $\beta_1$ by fitting the linear model:

$$y = \beta_0 + \beta_1 x + e \tag{13}$$

2. Using the sample standard deviation of the lowest concentration as an estimate for $\sigma_e$ and the standard deviation of the log of the replicates at the highest concentration as an initial estimate for $\sigma_\eta$, refit the model in step 1 using WLS with weights equal to

$$w(x) = 1/[\sigma_e^2 + \beta_1^2 x^2 e^{\sigma_\eta^2}(e^{\sigma_\eta^2} - 1)] \tag{14}$$

3. Using the new estimates of $\beta_0$ and $\beta_1$, compute the predicted response $\hat{y} = \hat{\beta}_0 + \hat{\beta}_1 x$ and standard error of the calibration curve at each concentration $x$:

$$s^2(x) = \frac{\sum_{i=1}^{m(x)} (\hat{y} - y_i)^2}{m(x)} \tag{15}$$

where $m(x)$ is the number of replicates for concentration $x$.

4. Using WLS, fit the variance function:

$$s^2(x) = \gamma + \delta x^2 + e \qquad (16)$$

where

$$\gamma = \sigma_e^2 \qquad (17)$$

and

$$\delta = \beta_1^2 e^{\sigma_\eta^2}(e^{\sigma_\eta^2} - 1) \qquad (18)$$

using weights:

$$w(x) = m(x)/s^2(x) \qquad (19)$$

5. Compute the new estimates of $\sigma_e^2 = \gamma$ and

$$\sigma_\eta^2 = \log_e\left[(1 + \sqrt{1 + 4\delta/\beta_1^2})/2\right] \qquad (20)$$

6. Iterate until convergence.

In general, this algorithm will converge to positive values of $\gamma$ and $\delta$. Note that this algorithm uses WLS to compute the parameters of the calibration curve ($\beta_0$ and $\beta_1$) as well as the parameters of the variance function ($\gamma$ and $\delta$). In this way, the lowest concentrations with the smallest variances provide the greatest weight in the estimation. The net result is to not sacrifice precision in estimating the calibration function and corresponding interval estimates at low levels by including higher concentrations in the analysis. This is quite useful if the interest is in low-level detection and quantification.

## Exponential Model

An alternative parameterization of the variance function involves modeling the relationship between $\sigma$ and $x$ as an exponential function of the following form:

$$\sigma_x = a_0 e^{a_1(x)} \qquad (21)$$

Although less well theoretically motivated than the Rocke and Lorenzato model, the exponential model provides excellent fit to a wide variety of analytical data (see Gibbons et al., 1997). The model can be applied either to the observed standard deviations at each concentration or, iteratively, to the sum of squared residuals. For estimating $a_0$ and $a_1$, the traditional approach involves substituting $s_x$ for $\sigma_x$ and using nonlinear least squares (e.g., Gauss-Newton) or using OLS regression of the natural log transformed observed standard deviation on true concentration (Snedecor and Cochran, 1989). Similarly, WLS can also be used on the regression of $\log_e(s)$ on $x$ using weights

$$w(x) = m(x)/s^2(x) \tag{22}$$

## Linear Model

The linear model has also been used to model the variance function (Currie, 1995). The linear model is of the form

$$\sigma_x = a_0 + a_1(x) \tag{23}$$

The primary disadvantage of the linear model is that the small sampling fluctuations in the observed sample variance at each concentration can lead to a negative intercept (i.e., $a_0 < 0$) and negative variance estimates. This can lead to improper detection and quantification limit estimates and corresponding interval estimates. As such, the linear model is generally not recommended for routine use. This is not a problem for either of the two preceding models, which can mimic a linear model if required.

## ITERATIVELY REWEIGHTED LEAST SQUARES ESTIMATION

An alternative to modeling the observed variance at each concentration is to model the squared residuals as a function of $x$, and then to use this estimated variance function to obtain weights that are then used in estimating the regression coefficients. This process is iterated until convergence, hence the term "iteratively reweighted least squares'" (Carroll and Rupert, 1988). As noted by Neter et al. (1990) the methods of maximum likelihood and weighted least squares lead to the same estimators for linear regression models of the form considered here. The previous example of the WLS estimator for the Rocke and Lorenzato model is an example of iteratively reweighted least squares. The general algorithm is as follows:

1. Use OLS estimation to find initial estimates of $\beta_0$ and $\beta_1$ by fitting the linear model

$$y = \beta_0 + \beta_1 x + \varepsilon \tag{24}$$

2. Using the OLS estimates of $\beta_0$ and $\beta_1$, compute the predicted response $\hat{y} = \hat{\beta}_0 + \hat{\beta}_1 x$ and the standard error of the calibration curve at each concentration $x$:

$$s^2(x) = \frac{\sum_{i=1}^{m(x)} (\hat{y} - y_i)^2}{m(x)} \qquad (25)$$

where $m(x)$ is the number of replicates for concentration $x$.

3. Using an appropriate model for the variance function, fit the variance function to the sum of squared residuals:

$$s^2(x) = f(x) \qquad (26)$$

4. Using the provisional weights:

$$w(x) = m(x)/s^2(x) \qquad (27)$$

recompute $\beta_0$ and $\beta_1$ using WLS.

5. Iterate until convergence.

## WLS PREDICTION INTERVALS

For WLS estimates of $\beta_0$ and $\beta_1$ the estimated variance for a predicted value $\hat{y}_{wj}$ is

$$V(\hat{y}_{wj}) = s_w^2 \left[ k_j + \frac{1}{\sum_{i=1}^{n}(1/k_i)} + \frac{(x_j - \bar{x}_w)^2}{\sum_{i=1}^{n}(x_i - \bar{x}_w)^2/k_i} \right] \qquad (28)$$

where $k_j$ is the estimated variance at concentration $x_j$. An upper $(1-\alpha)100$ percent confidence interval for $\hat{y}_{wj}$ (i.e., an upper prediction limit for a new measured concentration or instrument response at true concentration $x_j$) is

$$\hat{y}_{wj} + t\sqrt{V(\hat{y}_{wj})} \qquad (29)$$

where $t$ is the upper $(1-\alpha)100$ percent point of Student's $t$-distribution on $n-2$ degrees of freedom. For example, at $x=0$, the upper prediction limit (UPL) is

$$UPL = \frac{ts_w}{b_{1w}} \sqrt{s_0^2 + \frac{1}{\sum_{i=1}^{n}(1/k_i)} + \frac{(-\bar{x}_w)^2}{\sum_{i=1}^{n}(x_i-\bar{x}_w)^2/k_i}} \qquad (30)$$

where $s_0^2$ is the variance of the measured concentrations or instrument responses for a sample that does not contain the analyte.

## CONFIDENCE REGION FOR AN UNKNOWN TRUE CONCENTRATION

In general practice, measured concentrations are reported as if they are true concentrations, without the benefit of an index of uncertainty. There are two problems with this. First, the measured concentration may provide a biased estimate of the true concentration to the extent that $\beta_0 \neq 0$. Second, even in the absence of bias, the measured concentration is only an estimate of the true concentration, and it has a level of uncertainty that is ignored by simply presenting the measured concentration in the absence of a proper uncertainty interval.

To provide an estimate of true concentration $x$ from measured concentration $y$ for the Rocke and Lorenzato model, compute

$$\hat{x} = \frac{y - \hat{\beta}_0}{\hat{\beta}_1 \hat{\gamma}} \qquad (31)$$

Bhaumik and Gibbons (2005) derived the asymptotic variance of $\hat{x}$ as

$$Var(x) = \frac{\sigma_e^2}{\beta_1^2 \gamma^2}(1+1/n_0) + x^2(\gamma^2 - 1) \qquad (32)$$

where $n_0$ is the number of calibration measurements at or near zero. As expected, the variance of $\hat{x}$ depends on $x$ and increases with increasing concentration.

Bhaumik and Gibbons (2005) developed a confidence interval for an unknown true concentration $x$ given a measured concentration $y$, separately for true concentrations at or near $x=0$ and for larger non-zero true concentrations. For a low-level true concentration $x_0$, the $(1-\alpha)100$ percent confidence region for $x_0$ is $\max(0, y_0 - z_{\alpha/2}\sqrt{\hat{\sigma}_e^2/n_0}, y_0 + z_{\alpha/2}\sqrt{\hat{\sigma}_e^2/n_0})$. To construct a confidence interval for an unknown higher level concentration $x$, they use a lognormal approximation. Let

$$c_1 = Var(y) = \beta_1^2 x^2 (\gamma^4 - \gamma^2) + \sigma_e^2 \qquad (33)$$

$$c_2 = \frac{c_1}{\beta_1^2 x^2} \tag{34}$$

$$c_3 = ln\left(\frac{1+\sqrt{1+4c_2}}{2}\right) \tag{35}$$

where $c_3$ is the approximate variance of

$$ln\left(\frac{y-\beta_0}{\beta_1 x}\right) \tag{36}$$

The quantity

$$z(x) = \frac{ln(y-\beta_0) - ln(\beta_1 x)}{\sqrt{c_3}} \tag{37}$$

is distributed $N(0,1)$ so that the $(1-\alpha)100$ percent confidence region for $x_0$ is obtained by iteratively solving

$$\Re(x) = \{x : -z_{\alpha/2} \leq z(x) \leq z_{\alpha/2}\} \tag{38}$$

In addition to reporting measured concentrations, the point estimate of $x$ and its 95 percent confidence interval should also be routinely reported; it can be used for the purpose of making both detection decisions and comparisons to regulatory standards. If, for example, the lower 95 percent confidence limit is greater than zero, there is 95 percent confidence that the true concentration is greater than zero. By contrast, if the upper 95 percent confidence limit is less than a regulatory standard, there is 95 percent confidence that the true concentration is less than the regulatory standard, and the corresponding (and potentially less costly) disposal options can be pursued.

## DETECTION AND QUANTIFICATION

The previously described WLS prediction limit $\hat{y}_0$ corresponds to the concept of a decision limit $L_C$ defined by Currie (1968) for the case in which the data arise from a calibration experiment, $\mu$ and $\sigma$ at $x = 0$ are unknown, and one wishes to make a detection decision for a single future test sample. Measured concentrations (or instrument responses) that exceed the UPL should yield the binary decision of "detected" with $(1-\alpha)100$ percent confidence. Note that when the true concentration $x = L_C$, the probability of exceeding the UPL is only 50 percent. As such, Currie defined the detection limit $L_D$ as the 95 percent UPL for a true concentration at $L_C$. The WLS estimate of $L_C$ is therefore

$$L_C = \frac{ts_w}{b_{1w}} \sqrt{s_{L_C}^2 + \frac{1}{\sum_{i=1}^{n}(1/k_i)} + \frac{(L_C - \bar{x}_w)^2}{\sum_{i=1}^{n}(x_i - \bar{x}_w)^2/k_i}} \qquad (39)$$

and the WLS estimate of $L_D$ is

$$L_D = L_C + \frac{ts_w}{b_{1w}} \sqrt{s_{L_D}^2 + \frac{1}{\sum_{i=1}^{n}(1/k_i)} + \frac{(L_D - \bar{x}_w)^2}{\sum_{i=1}^{n}(x_i - \bar{x}_w)^2/k_i}} \qquad (40)$$

Note that in order to compute $L_C$ and $L_D$, one must have estimates of $s_{L_C}^2$ and $s_{L_D}^2$, which are often unavailable and must be estimated using a model of standard deviation versus concentration, as previously described. The final estimates of $L_C$ and $L_D$ are obtained from simple repeated substitution beginning from $L_C = 0$ and $L_D = L_C$ until convergence (i.e., change of less than $10^{-4}$ in estimates of $L_C$ and $L_D$ on successive iterations).

Finally, Currie (1968) defined the limit of determination $L_Q$ as the concentration at which the signal-to-noise ratio is 10 to 1. In the current context, one can estimate $L_Q$ by identifying the true concentration at which the estimated standard deviation is one-tenth of its magnitude. Again, a simple iterative approach generally performs quite well (Gibbons and Coleman, 2001; Gibbons et al., 1997).

## REFERENCES

Bhaumik, D. and R. Gibbons. 2005. Confidence regions for random-effects calibration curves with heteroscedastic errors. Technometrics 47(2): 223-231.

Carroll, R. and D. Rupert. 1988. Transformation and Weighting in Regression. Boca Raton, FL: CRC Press.

Currie, L. 1968. Limits for qualitative detection and quantitative determination: Application to radiochemistry. Analytical Chemistry 40(3): 586-593.

Currie, L. 1995. Nomenclature in evaluation of analytical methods including detection and quantification capabilities. Pure and Applied Chemistry 67: 1699-1723.

Gibbons, R. 1995. Some statistical and conceptual issues in the detection of low-level environmental pollutants. Environmental and Ecological Statistics 2(2): 125-145.

Gibbons, R. and D. Bhaumik. 2001. Weighted random-effects regression models with applications to interlaboratory calibration. Technometrics 43(2): 192-198.

Gibbons, R. and D. Coleman. 2001. Statistical Methods for Detection and Quantification of Environmental Contamination. New York, N.Y.: John Wiley & Sons, Inc.

Gibbons, R., D. Coleman, and R. Maddalone. 1997. An alternate minimum level definition for analytical quantification. Environmental Science & Technology 31(7): 2071-2077.

Neter, J., W. Wasserman, and M. Kutner. 1990. Applied Linear Regression Models, 2nd Ed. Homewood, IL: McGraw-Hill/Irwin.

Rocke, D. and B. Durbin. 1998. Models and Estimators for Analytical Measurement Methods with Non-constant Variance. Davis, Calif.: University of California at Davis, Center for Image Processing and Integrated Computing.

Rocke, D. and S. Lorenzato. 1995. A two-component model for measurement error in analytical chemistry. Technometrics 37(2): 176-184.

Snedecor, G. and W. Cochran. 1989. Statistical Methods. Ames, IA: Iowa University Press.

# Appendix E
# Sampling Variability and Uncertainty Analyses

In Appendix D, uncertainty in the analytical measurement process was considered and confidence intervals that reflect that uncertainty in an unknown true concentration $x$ were developed. However, if one obtains a series of $n$ measurements of a given piece of equipment, or of an area of potential contamination such as a room, or $n$ soil samples in an area where contamination may have occured, then inferences about the potential area of concern must incorporate the sampling variability associated with the $n$ measured concentrations. In a perfect world, one would compute a $(1-\alpha)100$ percent normal upper confidence limit (UCL), and if the UCL was less than the regulatory standard, one could conclude with $(1-\alpha)100$ percent certainty that the true concentration mean for the piece of equipment or spatial area was less than the regulatory standard of interest. Note that this does not require all measurements to be below the regulatory standard. Of course, the converse is also true—namely, that all of the individual measurements can be below the regulatory standard but the UCL may still exceed the standard. It should be noted that there is considerable EPA guidance supporting this approach, including but not limited to SW846 (EPA, 2007) guidance and the EPA unified statistical guidance document (EPA, 2009). In addition, this general approach is also clearly recommended in the ASTM consensus standard (D7048) (ASTM, 2010).

Factors that complicate the simple use of a normal UCL are these: (1) the distribution of measured concentrations is rarely normal and generally has a long right tail, which is characteristic of a lognormal or gamma distribution; (2) the analyte is often not detected in a substantial proportion of the samples; and (3) the large number of statistical comparisons that are made leads to a large number of positive results, consistent with chance expectations but likely to be false positives. In the following sections, a general statistical methodology that can be followed to address such factors is outlined.

## NORMAL CONFIDENCE LIMITS FOR THE MEAN[1]

For a normally distributed constituent that is detected in all cases the $(1-\alpha)100$ percent normal lower confidence level (LCL) (assessment sampling and monitoring) for the mean of $n$ measurements is computed as

---

[1] The remainder of this appendix is largely an adaptation from Gibbons, 2009.

$$\bar{x} - t_{[n-1,\alpha]} \frac{s}{\sqrt{n}} \qquad (1)$$

The $(1-\alpha)100$ percent normal UCL (corrective action) for the mean of $n$ measurements is computed as

$$\bar{x} + t_{[n-1,\alpha]} \frac{s}{\sqrt{n}} \qquad (2)$$

When nondetects are present, several reasonable options are possible. If $n < 8$, nondetects are replaced by one-half of the detection limit (DL) since with fewer than eight measurements, more sophisticated statistical adjustments are typically not appropriate. Similarly, a normal UCL is typically used because seven or fewer samples are insufficient to confidently determine distributional form of the data. Because of a lognormal limit with small samples can result in extreme limit estimates, it is reasonable and conservative to default to normality for cases in which $n < 8$.

If $n \geq 8$, a good choice is to use the method of Aitchison (1955) to adjust for nondetects and test for normality and lognormality of the data using the Shapiro-Wilk test. However, the ability of the Shapiro-Wilk test (and other distributional tests) to detect nonnormality is highly dependent on sample size. For most applications, 95 percent confidence is a reasonable choice. Note that alternatives such as the method of Cohen (1961) can be used; however, the DL must be constant.

## LOGNORMAL CONFIDENCE LIMITS FOR THE MEDIAN

For a lognormally distributed constituent—that is, $y = log_e(x)$ is distributed $N(\mu_y, \sigma_y^2)$—the $(1-\alpha)100$ percent LCL for the median or 50th percentile of the distribution is given by

$$\exp\left[\bar{y} - t_{[n-1,\alpha]} \frac{s_y}{\sqrt{n}}\right] \qquad (3)$$

where $\bar{y}$ and $s_y$ are the mean and standard deviation of the natural log transformed concentrations. Note that the exponentiated limit is, in fact, an LCL for the median and not the mean concentration. In general, the median and corresponding LCL will be lower than the mean and its corresponding LCL. The $(1-\alpha)100$ percent UCL for the median or 50th percentile of the distribution is given by

$$\exp\left[\bar{y} + t_{[n-1,\alpha]} \frac{s_y}{\sqrt{n}}\right] \qquad (4)$$

# LOGNORMAL CONFIDENCE LIMITS FOR THE MEAN

## The Exact Method

Land (1971) developed an exact method for computing confidence limits for linear functions of the normal mean and variance. The classic example is the normalization of a lognormally distributed random variable $x$ through the transformation $y = log_e(x)$, where, as noted previously, $y$ is distributed normal with mean $\mu$ and variance $\sigma^2$, or $y: N(\mu_y, \sigma_y^2)$. Using Land's (1975) tabled coefficients $H_\alpha$, the one-sided $(1-\alpha)100$ percent lognormal LCL for the mean is

$$\exp\left(\bar{y} + .5s_y + \frac{H_\alpha s_y}{\sqrt{n-1}}\right) \tag{5}$$

Alternatively, using $H_{1-\alpha}$, the one-sided $(1-\alpha)100$ percent lognormal UCL for the mean is

$$\exp\left(\bar{y} + .5s_y + \frac{H_{1-\alpha} s_y}{\sqrt{n-1}}\right) \tag{6}$$

The factors $H$ are given by Land (1975) and $\bar{y}$ and $s_y$ are the mean and standard deviation of the natural log transformed data (i.e., $y = log_e(x)$). Gilbert (1987) has a small subset of these extensive tables for $n = 3$ through 101, $s_y = .1$ through 10.0, and $\alpha = .05$ and .10 ( i.e., upper and lower 90 percent and 95 percent confidence limit factors). Because these tables had historically been difficult to find, Gibbons and Coleman (2001) reproduced the complete set of Land's (1975) tables and have also included computing approximations that can be used for automated applications. Land (1975) suggests that cubic interpolation (i.e., four-point Lagrangian interpolation) be used when working with these tables (Abramawitz and Stegun, 1964). A much easier and quite reasonable alternative is to use logarithmic interpolation.

## Approximate Lognormal Confidence Limit Methods

There are also several approximations to lognormal confidence limits for the mean that have been proposed. These have been conveniently classified as either transformation methods or direct methods (Land, 1970). A transformation method is one in which the confidence limit is obtained for the expected value of some function of $x$ and then transformed by some appropriate function to give an approximate limit for the expectation of $x$ (i.e., $E(x)$), which in the lognormal case is $E(x) = \mu + \frac{1}{2}\sigma^2$. This estimate is assumed to be normally distributed and approximate confidence limits are computed accordingly.

The simplest transformation method is the naive transformation, which simply involves taking a log transformation of the data, computing the confidence limit on a log scale, and then exponentiating the limit. As previously noted, this is, in fact, a confidence limit for the median and not the mean. The method provides somewhat reasonable results as a confidence limit for the mean when $\sigma_y$ is very small but deteriorates quickly as $\sigma_y$ increases (Land, 1970).

Patterson (1966) proposed use of the transformation

$$\hat{\mu}_x = \exp\left(\bar{y} + \frac{1}{2}\sigma_y^2\right) \tag{7}$$

to remove the obvious bias of the naive method. Patterson's transformation would be exact if $\sigma_x^2$ were known; however, when the variance is unknown, it too behaves poorly when $\sigma_y$ increases (Land, 1970). More complicated alternatives described by Finney (1941) and Hoyle (1968) provide results similar to those of Patterson's transformation and are therefore not presented.

Direct methods offer an advantage over transformation methods in that they obtain confidence intervals directly for $E(x)$ or some function of $E(x)$. In light of this, these methods do not suffer from the bias introduced by failing to take into account the dependence of $E(x)$ on both $\mu$ and $\sigma^2$. However, by applying normality assumptions to $E(x)$, direct estimates can produce inadmissible confidence limits for $E(x)$. To this end, Aitchison and Brown (1957) have suggested computing the usual normal confidence limit, which under the Central Limit theorem should converge to exact limits as $n$ becomes large. Hoyle (1968) suggested replacing $\bar{x}$ and $s_x^2/n$ by their minimum variance unbiased estimates (MVUE). Finney (1941) derived the MVUE of $E(x)$ as follows

$$\hat{\theta} = \exp(\bar{y})\psi((1-n^{-1})s_y^2) \tag{8}$$

and Hoyle (1968) derived the MVUE for the variance of $E(x)$ as

$$\hat{\varphi} = \exp(2\bar{y})\left\{\psi^2\left[(1-m^{-1})s_y^2\right] - \psi\left[(2-4n^{-1})s_y^2\right]\right\} \tag{9}$$

where

$$\psi(g) = 1 + \frac{n-1}{n}g + \frac{(n-1)^3}{n^2 2!}\frac{g^2}{n+1} + \frac{(n-1)^5}{n^3 3!}\frac{g^3}{(n+1)(n+3)} + \ldots \tag{10}$$

is a Bessel function with argument $g$. In this method, the normal quantile $z_\alpha$ replaces $t_{n-1,\alpha}$ since there is no reason to believe that $\hat{\phi}$ is chi-squared and independent of $\hat{\theta}$. Unfortunately, Land (1970) has shown that these methods are only useful for large $n$ (i.e., $n > 100$) and even there only for small values of $s_y$.

The final direct method, which is attributed to D.R. Cox, has been shown to give the best overall results of any of the approximate methods (Land, 1970). The MVUE of $\beta = \log E(x)$ is $\hat{\beta} = \bar{y} + 1/2 s_y^2$, and the MVUE of the variance $\gamma^2$ of $\hat{\beta}$ is

$$\hat{\gamma}^2 = s_y^2/n + \frac{1}{2} s_y^4/(n+1) \tag{11}$$

Assuming approximate normality for $\hat{\beta}$, one may obtain approximate confidence limits for $E(x)$ of the form

$$\text{LCL} = \exp(\hat{\beta} - z_\alpha \hat{\gamma}) \tag{12}$$

and

$$\text{UCL} = \exp(\hat{\beta} + z_\alpha \hat{\gamma}) \tag{13}$$

## NONPARAMETRIC CONFIDENCE LIMITS FOR THE MEDIAN

When data are neither normally or lognormally distributed or the detection frequency is too low (e.g., < 50%) for a meaningful distributional analysis, nonparametric confidence limits become the method of choice. The nonparametric confidence limit is defined by an order statistic (i.e., a ranked observation) of the $n$ measurements. Note that in the nonparametric case, one is restricted to computing confidence limits on percentiles of the distribution, for example, the 50th percentile or median of the on-site/downgradient distribution. Unless the distribution is symmetric (i.e., the mean and median are equivalent), there is no direct nonparametric way of constructing a confidence limit for the mean concentration.

To construct a confidence limit for the median concentration, one uses the fact that the number of samples falling below the $p(100)$th percentile of the distribution (e.g., $p = .5$, where $p$ is between 0 and 1) out of a set on $n$ samples will follow a binomial distribution with parameters $n$ and success probability $p$, where success is defined as the event that a sample measurement is below the $p(100)$th percentile. The cumulative binomial distribution, $\text{Bin}(x; n, p)$, represents the probability of getting $x$ or fewer successes in $n$ trials with success probability $p$, and can be evaluated as

$$\text{Bin}(x; n, p) = \sum_{i=1}^{x} \binom{n}{i} p^i (1-p)^{n-i} \tag{14}$$

The notation $\binom{n}{i}$ denotes the number of combinations of $n$ things taken $i$ at a time, where

$$\binom{n}{i} = \frac{n!}{i!(n-i)!} \tag{15}$$

and $k! = 1 \cdot 2 \cdot 3 \ldots k$ for any counting number $k$. For example, the number of ways in which two things can be selected from three things is

$$\binom{3}{2} = \frac{3!}{2!(1)!} = \frac{1 \cdot 2 \cdot 3}{(1 \cdot 2)(1)} = \frac{6}{2} = 3 \tag{16}$$

To compute a nonparametric confidence limit for the median, begin by rank ordering the $n$ measurements from smallest to largest as $x_{(1)}, x_{(2)}, \ldots, x_{(n)}$. Denote the candidate end points selected to bracket the 50th percentile (i.e., $(n+1)*.5$) as $L^*$ and $U^*$ for lower and upper bound, respectively. For the LCL, compute the probability

$$1 - \text{Bin}(L^* - 1; n, .5) \tag{17}$$

If the probability is less than the desired confidence level, $1-\alpha$, select a new value of $L^* = L^* - 1$ and repeat the process until the desired confidence level is achieved. For the UCL, compute the probability

$$1 - \text{Bin}(U^* - 1; n, .5) \tag{18}$$

If the probability is less than the desired confidence level, $1-\alpha$, select a new value of $U^* = U^* + 1$ and repeat the process until the desired confidence level is achieved. If the desired confidence level cannot be achieved, set the LCL to the smallest value or the UCL to the largest value and report the achieved confidence level.

Another distribution that is often used for skewed data is the gamma distribution. Suppose $x$ follows a gamma distribution with the shape parameter $\kappa$ and scale parameter $\theta$. Then the gamma density is given by

$$f(x) = \frac{1}{\Gamma(\kappa)\theta^\kappa} x^{\kappa-1} e^{-\frac{x}{\theta}} \tag{19}$$

Let $x_{(1)}, x_{(2)}, \ldots, x_{(n)}$ be a random sample of size $n$ drawn from this population to estimate the unknown parameters. Denote the arithmetic and geometric means based on this random sample by $\bar{x}$ and $\tilde{x}$, respectively. The maximum likelihood estimators of $\theta$ and $\kappa$, denoted by $\hat{\theta}$ and $\hat{\kappa}$, are solutions to the following equations:

$$ln(\hat{\kappa}) - \psi(\hat{\kappa}) = \ln(\bar{x}/\tilde{x}), \text{ and } \hat{\kappa}\hat{\theta} = \bar{x} \quad (20)$$

where $\psi$ denotes a digamma or Euler's psi function. The mean and variance of $x$ are:

$$E(x) = \kappa\theta \text{ and } V(x) = \kappa\theta^2 \quad (21)$$

To construct the UCL for this type of data, Aryal et al. (2009) constructed the following statistic:

$$T = \frac{9(n\mu)^{1/3}(n-1)(X^{1/3} - (n\mu)^{1/3})^2}{2n\mu R_n} \quad (22)$$

where $R_n$ is the logarithm of the ratio of the arithmetic mean to the geometric mean and $\mu$ is the mean of the population. $X$ is the sum of all the observations. The UCL of $\mu$ is obtained by solving the following equation and taking the largest root:

$$T \leq F_{1-\alpha, 1, n-1} \quad (23)$$

where $F_{1-\alpha}$ is the $(1-\alpha)$ 100th percentile of the $F$ distribution with degrees of freedom 1 and $n-1$. To compute the $(1-\alpha)100$ percent UCL, invert the test statistic $T$, from which one obtains

$$\text{UCL} = \frac{\sum x_i}{n(1-\sqrt{U})^3} \quad (24)$$

where

$$U = 2\ln\left(\frac{\bar{x}}{\tilde{x}}\right)\frac{F_{1-\alpha}}{9(n-1)} \quad (25)$$

## REFERENCES

Abramawitz, M. and I. Stegun. 1964. Handbook of Mathematical Functions with Formulas, Graphs, and Mathematical Tables. Washington, D.C.: National Bureau of Standards.

Aitchison, J. 1955. On the distribution of a positive random variable having a discrete probability mass at the origin. Journal of American Statistical Association 50: 901-908.

Aitchison, J. and J. Brown. 1957. The Log-normal Distribution. Cambridge, UK: Cambridge University Press.

Aryal, S., D. Bhaumik, S. Santra, and R. Gibbons. 2009. Confidence interval for random-effects calibration curves with left-censored data. Environmetrics 20(2): 181-189.

ASTM (American Society for Testing and Materials). 2010. ASTM D7048-04 Standard Guide for Applying Statistical Methods for Assessment and Corrective Action Environmental Monitoring Programs. West Conshohocken, Pa.: ASTM International.

Cohen, A. 1961. Tables for maximum likelihood estimates: singly truncated and singly censored samples. Technometrics 3: 535-541.

U.S. Environmental Protection Agency (EPA). 2007. SW-846 Test Methods for Evaluating Solid Waste, Physical/Chemical Methods. Washington, D.C.: Environmental Protection Agency.

EPA. 2009. Statistical Analysis of Groundwater Monitoring Data at RCRA Facilities Unified Guidance. EPA 530/R-09-007. Washington, D.C.: Environmental Protection Agency Office of Resource Conservation and Recovery.

Finney, D. 1941. On the distribution of a variate whose logarithm is normally distributed. Journal of the Royal Statistical Society, Series B 7: 155-161.

Gibbons, R. 2009. Assessment and corrective action monitoring. Pp. 317-335 in Statistical Methods for Groundwater Monitoring, edited by R. Gibbons, D. Bhaumik, and S. Aryal. Hoboken, N.J.: John Wiley & Sons, Inc.

Gibbons, R. and D. Coleman. 2001. Statistical Methods for Detection and Quantification of Environmental Contamination. New York, N.Y.: John Wiley & Sons, Inc.

Gilbert, R. 1987. Statistical Methods for Environmental Pollution Monitoring. New York, N.Y.: John Wiley and Sons, Inc.

Hoyle, M. 1968. The estimation of variances after using a gaussianating transformation. Annals of Mathematical Statistics 39: 1125-1143.

Land, L. 1970. Phreatic Versus Vadose Meteoric Diagenesis of Limestones: Evidence from a Fossil Water Table.

Land, C. 1971. Confidence intervals for linear functions of the normal mean and variance. Annals of Mathematical Statistics 42:1187-1205.

Land, C. 1975. Tables of confidence limits for linear functions of the normal mean and variance. Selected Tables in Mathematical Statistics 3: 385-419.

Patterson, C. and D. Settle. 1966. 7th Materials Research Symposium. National Bureau of Standards Special Publication 422. Washington, D.C.: U.S. Government Printing Office.